CHANGING CONSUMERS AND CHANGING TECHNOLOGY IN HEALTH CARE AND HEALTH CARE DELIVERY

RESEARCH IN THE SOCIOLOGY OF HEALTH CARE

Series Editor: Jennie Jacobs Kronenfeld

RESEARCH IN THE SOCIOLOGY OF HEALTH CARE
VOLUME 19

CHANGING CONSUMERS AND CHANGING TECHNOLOGY IN HEALTH CARE AND HEALTH CARE DELIVERY

EDITED BY

JENNIE JACOBS KRONENFELD

Department of Sociology, Arizona State University, USA

2001

JAI
An Imprint of Elsevier Science

Amsterdam – London – New York – Oxford – Paris – Shannon – Tokyo

ELSEVIER SCIENCE Ltd
The Boulevard, Langford Lane
Kidlington, Oxford OX5 1GB, UK

First edition 2001

Library of Congress Cataloging in Publication Data
A catalog record from the Library of Congress has been applied for.

British Library Cataloguing in Publication Data
A catalogue record from the British Library has been applied for.

ISBN: 0-7623-0808-7
ISSN: 0275-4959 (Series)

♾The paper used in this publication meets the requirements of ANSI/NISO Z39.48-1992 (Permanence of Paper).
Printed in The Netherlands.

CONTENTS

LIST OF CONTRIBUTORS

Letitia T. Alston	Institute for Science, Technology and Public Policy Texas A&M University, USA
Caryn Aviv	Breast Care Center University of California, USA
Peri J. Ballantyne	Department of Public Health Science University of Toronto, Canada
Jeff Belkora	Consultation Planning Services Community Breast Health Project Palo Alto, USA
John R. Booher	Texas Instruments, USA
Neale R. Chumbler	North Florida/South Georgia Veterans Health System, and the University of Florida, USA
Laura Esserman	Community Breast Health Project University of California, USA
Beth Furlong	School of Nursing Creighton University, USA
James W. Grimm	Department of Sociology Western Kentucky University, USA
Jeffrey Hall	Department of Sociology, University of Alabama, Birmingham, USA
Barbara Hanson	Department of Sociology York University, Canada
Ted Hartman	Blue Cross/Blue Shield Texas, USA

Gillian A. Hawker	Faculty of Medicine University of Toronto, Canada
Eva Kahana	Department of Sociology Case Western Reserve University, USA
Boaz Kahana	Department of Psychology Cleveland State University, USA
Jennie Jacobs Kronenfeld	Department of Sociology Arizona State University, USA
William McCaughan	School of Distance and Continuing Education. Oregon State University, USA
William Alex McIntosh	Institute for Science, Technology and Public Policy, Texas A&M University, USA
Eliza K. Pavalko	Department of Sociology Indiana University, USA
Detelina Radoeva	Institute for Human Development Life Course and Aging, University of Toronto, Canada
Susan C. Reed	DePaul University, Chicago, USA
Clasina Segura	Department of Sociology Texas A&M University, USA
Dianne Sykes	Programme Director, Human Relations/Sociology, Marian College, USA
Karen Sepucha	Foundation for Informed Medical Decision Making, Boston, USA
Eleanor V. Toney	Department of Sociology University of Georgia, USA
Robert R. Weaver	Department of Sociology and Anthropology, Youngstown State University, USA
Regina E. Werum	Sociology Department Emory University, USA

E. Jay Wheeler Department of Health Services Research and Management, Texas Tech. Health Sciences Center, USA

Marlene Wilken School of Nursing Creighton University, USA

Terri A. Winnick Department of Sociology Indiana University, USA

William C. Yoels Department of Sociology University of Alabama Brimingham, USA

INTRODUCTION

Jennie Jacobs Kronenfeld

This volume explores the roles of changing consumers and changing technology in health care and health care delivery today. Most of the focus is on the U.S. health care system in this volume; however, as is discussed in the following paragraphs, many of the issues raised are relevant for health care systems in a variety of countries. Some of the articles focus on special aspects of how new technologies and programs apply to either general groups within the health care system (patients, certain types of doctors) or more specialized groups, such as people with a certain health care problem. Papers deal with a wide variety of topics, from a focus on consumers and the varying roles they play in the emerging and changing U.S. health care system to the examination of specific principles such as social network approaches and how they can be applied to the examination of patient-provider roles to papers that incorporate the application of specific technological innovations within medicine, such as the impact of telemedicine or of computer tools and knowledge coupling on the delivery of health care. One factor that seems most true of the U.S. health care system (and many would argue of health care systems across the world) is that such systems are having to respond to changes in technology within health care, changes in technology not specific to health care, and changes in the way patients and physicians view health and view the use of services in society. In a variety of ways, each of these papers contributes something to these types of debates. As most of the papers are written by medical sociologists and part of that research tradition as well, they also contribute to an understanding of theoretical concerns within medical sociology, and the ways that studies of actual situations within health care can be used to broaden our understanding of society more generally. In addition, once researchers begin to consider issues of health care delivery, debates about the role of government and changes in those roles and trends also become important. While 50 years ago, many articles could focus on issues in medical sociology and even health care more broadly and not consider the role of government as a provider or payer of services or as an organizer of services, in today's health care system, these types of issues end up being discussed even in papers that focus on community-based studies, historical studies, or issues with patients in specific types of care settings.

Health policy changes are sweeping around the globe. Widely held simpler notions focused on market oriented reforms versus state oriented reforms are being rejected by social scientists who study health care and by policy researchers, as has been discussed in detail in a number of recent articles in the social sciences and health policy literature (Gauld, 2000; Peterson, 2000). The last decade has seen the growth of managed care within the U.S. and what some have termed a "market transformation of the U.S. health care system" (Brown, 1998). These trends have led to research on topics such as competition among health care plans, studies of provider incentive structures, studies on the role of consumer choice, and studies on plan contracting and its implications for physicians and consumers. By the late 1990s, journals were publishing papers on the rise of market thinking in other countries also, even in countries in which such approaches were very different from the ongoing thrust of the health care system such as New Zealand or Great Britain (Gauld, 2000). At the same time that other countries were looking at the growth of managed care in the U.S. and its market-dominated approach as an interesting potential solution to their own counties' health care problems, problems with managed care were beginning to emerge and some of the negative aspects of this approach were discussed in newspapers, and in the media more generally, as well as by scholars.

In the last few years, there have been a number of studies skeptical of the dominance of market-oriented approaches to care, some from an economic and some from a political science approach (Rice, 1998; Stone, 1998). This volume includes a number of articles that raise these questions as well, but from a more clearly sociological perspective, as well as from a perspective that incorporates the overall theme of this volume, changing consumers and changing technology and their impact upon the current health care system. There has been discussion of a "managed care backlash" resulting in ongoing debates within the U.S. Congress and in many state legislatures about whether a patients' bill of rights needs to be passed. While such legislation has not yet passed at a national level, a number of states over the last five years have passed portions of such legislation. From a sociological perspective, Gray (1997) has argued that trust is essential to the doctor/patient relationship and that a system based on trust in the competence and fiduciary ethic of individual physicians is being replaced by a system based on alternatives to trust. A number of papers in this volume relate to some of these types of issues.

Other challenges are also again being discussed in U.S. health care circles. After close to a decade of some success in controlling the costs of health care, fears about sky rocketing health care expenditures are again surfacing, both in the popular press and also in scholarly journals (Blumenthal, 2001). The Health Care Financing Administration (HCFA, 2001) reports that national health care expenditures have been growing at an average rate of 6.5% annually from 1998 through 2001, versus only 5.0% from 1993 through 1996. This is consistent with reports by Hogan and colleagues (Hogan, Ginsberg & Gabel, 2000) that private expenditures for increased by 6.6% per insured person in 1999, up from 5.1% in 1998 and 3.1% in 1997. Some sources report that HCFA will soon revise upward its projections of future health care spending, and that the current 13.6% of Gross Domestic Product that the U.S. spends on health care each year may grow to 16% by 2008 and perhaps as high as 25% by 2030 (Pear, 2000; Morris, 2000).

Some experts believe that the success in controlling health care costs in the early 1990s was due to the growth of managed care and its savings that were achieved through negotiated price discounts from physicians and other health care providers. The growing resistance from providers and patients about managed care, as discussed in current literature and in some of the selections in this volume, is one factor leading to the increase in health care expenditures. Another factor beginning to concern health policy experts is fundamental pressures within the health care system such as the cost of new technology and the availability of new diagnostic and therapeutic approaches (Blumenthal, 2001). Papers in this volume deal with both of these important current concerns in health care, the debate about managed care and its role in the U.S. health care system, and the role of technology in a changing health care system.

The papers in Volume 19 are divided into four sections. The first section includes four papers on Changing Consumer Perspectives in Today's Health Care Environment. The first paper in this section, by Beth Furlong and Marlene Wilken, explores the issue of managed care and its changing environment for consumers and health care providers. The authors argue (and I would agree) that managed care has become the predominant model of health care in the U.S. at this point in time. Given that, understanding more about managed care is a most appropriate beginning for this book. The paper includes a useful review of research literature on the strengths and weaknesses of the managed care system, and continues with a discussion of the problems that managed care has created and the state and federal regulatory legislation and litigation resulting from managed care. The second and third papers in this section continue with a focus on consumers, and explore how proactive health care consumers can make the health care system work for them, and how consumer desire interacts with the practice of medicine. The fourth paper in this section applies social network principles to the linkage between providers and patients.

The paper by the Kahanas "On Being a Proactive Health Care Consumer: Making an 'Unresponsive' System Work for You" is an excellent follow-up to the exploration of managed care. The paper argues that the ideals of patient centered care are increasingly advocated in health services research literature, but that for a patient to optimize his or her own care given the complexity of today's health care system and the lack of responsiveness often found, a patient needs to develop proactive strategies. These include improved gathering of information, using communication skills, and finding people behind the system to cooperate with the patient. Hansen presents a new theory of how the practice of medicine interacts with desires of consumers' medicalism. This is a general mode of thinking about both everyday life and social policy that explores the values consumers place on good health and ways in which medicine can become a "dumping ground" for many social problems. The last paper uses general principles of social networks such as global structure, structural equivalence, structural conductiveness, and duality of network linkages to better understand the evolving complexities of the interconnectedness among health care providers and patients. Thus this paper provides a more formalized and theoretical examination of topics discussed by the other papers in this section. Section Two includes three papers that explore specific programs and specific technologies in health care currently. Toney's paper is an analysis of hospital-technology interplay and its role as a determinant of mortality. The ecology theory of organizations serves as the underlying framework of this article. The paper explores how hospital services and their technology affect a community, and presents a methodologically sophisticated model for this process. Factors such as socioeconomic status, the presence of teaching hospitals and age of the population are all important factors. Reed's paper uses a very different methodology to explore issues of health care, specific programs and impact on the community. Using both historical data and a spatial analysis approach, the paper examines the effect of Medicaid upon the supply of long-term care facilities in Chicago communities and special impacts for the African American community. The paper closes by discussing the policy implications of these trends. Weaver's paper integrates the themes of this volume well by exploring two parallel developments in health care: the demand for greater patient participation in health care and the evolution of computer tools that help to inform patients of health care options. The impact of knowledge coupling computer tools in primary care is the specific focus of this paper.

The third section includes two papers that apply the issue of changing technology to providers of care. The first paper in this section, by McIntosh and colleagues, explores the impact of telemedicine. They point out that proponents of this type of technology believe it will resolve many of the problems associated with the lack of access to specialists for more isolated populations such as those in rural areas. For this to occur, physicians must be willing to use the technology, and this paper develops prediction models of continued use of this technology by primary care providers and by specialists. Characteristics such as age, gender, and years since graduation are important predictor factors. Winnick and colleagues examine a different aspect of health care, mental health education for family doctors. Trends in mental health education as presented in the family practice literature is the starting point for this examination. The impact of the development of the DSM and the DSM III is discussed, and linked with the development of drug treatment approaches.

The last section deals in some ways with a more narrow focus, issues of changes for specialized patients, especially those with serious illnesses. Aviv and colleagues use action research to improve collaboration between breast cancer patients and physicians. At a time when public attention on breast cancer has grown, this article discusses innovations in the care process. In addition to the exploration of breast cancer, the concluding two papers in the volume focus upon persons with traumatic injuries and arthritis patients. Hall and Yoels explore the subjective health status of three distinct groups of patients: those with spinal cord injuries, those with traumatic brain injuries, and those with burns and intra-articular fractures of the lower extremities. They argue that persons in each of these groups should be considered distinct with respect to the individual determinants of overall health status and the variable domains that explain the largest amount of variance in health status assessments. Ballantyne and colleagues expand upon an epidemiological survey that had found a discrepancy between medically assessed need and patient willingness to consider treatment that involved total joint arthroplasty. This qualitative study focuses upon the quality of the marital relationship, and argues that in addition to the individual's functional capacity, the couple's relational and functional behaviors influence the interpretation of the disease and the decisions made about its treatment. This paper provides a useful closing piece, reminding medical sociologists of the importance not only of technology, specific programs in health care, and changes in the delivery of health care, but of the social and structural factors such as marriage and quality of the marital relationship that form such an important part of the sociological world view.

REFERENCES

Blumenthal, D. (2001). Controlling Health Care Expenditures. *New England Journal of Medicine*, *344*, 766–769.

Brown, L. D. (1998). Exceptionalism As the Rule: U.S. Health Policy Innovation and Cross-National Learning. *Journal of Health Politics, Policy and Law*, *23*, 35–51.

Gauld, R. D. C. (2000). Big Bang and the Policy Prescription. *Journal of Health Policy, Politics and Law*, *25*, 815–844.

Gray, B. (1997). Trust and Trustworthy Care in the Managed Care Era. *Health Affairs*, *16*(January/Februrary), 34–54.

HCFA (2001). Website: http://www.hcfa/gov/stats/NHE/proj

Hogan, C., Ginsberg, P. B., & Gabel, J. R. (2000). Tracking Health Care Costs: Inflation Returns. *Health Affairs*, *19*, 217–223.

Morris, C. R. (2000). *Too Much of a Good Thing: Why Health Care Spending Won't Make us Sick*. New York: Century Foundation Press.

Pear, R. (2000). Health Costs Underestimated, Experts Say. *New York Times*, (November 30), A1.

Peterson, M. A. (2000). Is There a Future in Your Market? *Journal of Health Policy, Politics and Law*, *25*, 807–813.

Rice, T. (1998). *The Economics of Health Reconsidered*. Chicago: Health Administration Press.

Stone, D. (1998). Managed Care and the Second Great Transformation. *Journal of Health Policy, Politics and Law*, *24*, 1213–1218.

PART I:
CHANGING CONSUMER PERSPECTIVES IN TODAY'S HEALTH CARE ENVIRONMENT

MANAGED CARE: THE CHANGING ENVIRONMENT FOR CONSUMERS AND HEALTH CARE PROVIDERS

Beth Furlong and Marlene Wilken

ABSTRACT

"Managed care is the health law issue of 1999" (Leibold, 1999, p. 5). This chapter addresses changes in the health care system initiated by the managed care paradigm – which, in turn, has prompted many legal changes and challenges. Specifically, this chapter will describe the following: (a) an overview of the managed care system versus the traditional fee-for-service system; (b) a selected review of research literature on the strengths and limitations of the managed care system; and (c) a projection of the future trends of the managed care system for patients, consumers, health providers and society. The second area, the review of research literature, will include the impact of managed care on patients, consumers, physicians and nurses. In this chapter, the word patient, means someone who is ill; the word consumer, means someone who has health insurance and may become ill in the future and needs to utilize the health delivery system.

Changing Consumers and Changing Technology in Health Care and Health Care Delivery,
Volume 19, pages 3–20.
Copyright © 2001 by Elsevier Science Ltd.
All rights of reproduction in any form reserved.
ISBN: 0-7623-0808-7

DESCRIPTION OF THE TWO SYSTEMS OF HEALTH CARE DELIVERY

Managed care came about in large part as a response to the rapidly rising medical costs in a system widely believed to be inefficient. Large employers sought to reduce the rising costs of health insurance for their employees. One way to do this was to tell the workers that if they wanted higher wages they would have to agree to contracts offered by managed care companies that held down their health care benefits.

A managed care organization is "an entity that assumes both the clinical and financial responsibility for the provision of health care for a defined population" (Donaldson, 1998, p. 713). Managed care is a mechanism which puts economic constraints on the health care system (Riggs, 1996). In this system, a third party pays an agreed upon amount of money and health providers agree to provide certain health services. The change of the health care system from a fee-for-service to a managed care system is a recent phenomenon, which has greatly accelerated since 1985. At that time most Americans were covered by fee-for-service indemnity plans (Preuss, Committing to Care, 1998). Today, it is estimated that 65% of Americans are covered by the managed care system (Chin & Harrigill, 1999). An advantage of the managed care model is that it is 20–40% lower in cost than the fee-for-service system (Chin & Harrigill, 1999).

> One result of this trend toward managed care and HMOs has been a decrease in overall health service utilization over the past decade. One study finds that HMOs that employ their own physicians have reduced health service utilization by nearly 20% ... (Preuss, 1998, p. 7).

It is hypothesized such practice behavior by some physicians then spills over and influences other practice patterns in that geographic area. In the decade between 1985 and 1995, these decreases have been found – 12% decline in hospital discharges, 17% decrease in the hospital length of stay and 38% drop in the days of care per 1000 people (Preuss, 1998). In addition to the managed care model, many other changes have occurred – increased mergers of health systems, downsizing in health organizations, increased collective bargaining activity, increased integrated systems, and so forth.

REVIEW OF LITERATURE

Consumer

A 1998 Harris poll reflected that 50% of the American public thinks the health care system is getting worse. However, a different study showed 38% of the population approved of managed care (Friedman, 1999). A third study in 1998, carried out by the Employee Benefit Research Institute, found 73% of Americans believed they would only have access to care with significant financial hardship (Friedman, 1999).

In November of 1998, President Clinton asked Congress for a "Patient's Bill of Rights" which would offer a wide variety of protections to the more than 150 million Americans in managed care. Both the Democrats and Republicans introduced legislation after the new year, 1999. Both plans include the following: direct access for women to their obstetrician and gynecologist, emergency-room access guarantee without prior HMO approval, safeguard of personal information, allow doctors to discuss full range of medical options with patients and provide the right to appeal a managed-care decision by going to an outsider. The Democrat's plan allows for other patient's rights which include the right to sue health plans for improperly denying coverage, access to specialists, coverage for reconstructive breast surgery after mastectomy and continued treatment if the doctor is unexpectedly dropped from the health plan. The Republican plan adds provisions for expanded medical savings accounts and the creation of association health plans and regional supermarkets called HealthMarts to help both small businesses and families' shop for health insurance (Tumulty & Dickerson, 1998). Policy continues to be debated on this proposed bill in Spring 2001 following the inauguration of President Bush.

During 1998/99 several public opinion polls related to managed care were conducted by the Roper Center at the University of Connecticut. The majority of these surveys had over 1,000 participants. The topics included managed care or HMO, patient rights, or consumer rights. When 1,909 persons were given an array of issues which included protecting patient's rights, and asked which, if any, would be the deciding factor in their vote for Congress in 1998, only 2% indicated patient's rights, with 19% saying improving education, and 8% saying dealing with crime, keeping taxes down, and dealing with the economy and jobs (Public opinion on line, Question 9, 1998). When the public was asked to rate how serious they felt HMO's rules were that restrict the patient's rights to sue for damages if improperly denied care, 46% indicated it was extremely serious and 29% responded it was somewhat serious (Public opinion on line, Question 123, 1998).

Two separate polls were conducted during the same time span in July 1998. Both dealt with similar issues. In one poll, the questions were phrased by asking about the need for congressional action while the other poll phrased the questions by stating "some people think" and then asked with whom do you agree more. The question in one poll was as follows: Which of the following statements comes closer to your view about regulating HMO's (Health

Maintenance Organizations) and other managed care plans? Statement A: Congress should pass a Patient's bill of rights, because it is necessary to guarantee consumers more rights in dealing with their insurance plans, including the rights to appeal HMO's decisions and to have access to emergency rooms without advanced permission. Statement B: Congress should not pass a Patient's bill of rights, because it would increase the size of the federal bureaucracy and raise the cost of health insurance, which would result in some people being unable to afford coverage. Results indicated that 69% felt Congress should pass a Patient's bill of rights and 21% indicated no, with 6% not sure (Public opinion on line, Question 036, 1998).

In the other poll, the question read as follows: Some people think that patients should have the right to sue their HMOs for improper care, to prevent HMOs from only looking at costs when making medical decisions. Others think that giving patients the right to sue their HMO for improper care would just raise the cost of health insurance, which would result in some people no longer being able to afford coverage. With whom do you agree more? Those believing in the right to sue HMOs constituted 61% versus 29% who said no right to sue (Public opinion on line, Question 039, 1998). During this same time period, the public was asked "if a Patient's bill of rights were passed, how much do you think it would improve the medical care of people who are covered by HMOs – a great deal, quite a bit, just some, or very little?" Results were 24% a great deal, 24% quite a bit, 33% just some, and 13% very little (Public opinion on line, Question 040, 1998).

As the fall election drew closer, other Roper polls were conducted which included the Patient's bill of rights and ideas of partisan politics. One poll asked about a choice between Candidate A who favored a Patient bill of rights that guarantees the right to sue HMOs for improper care but might result in higher premiums, and Candidate B who favors a Patient bill of rights that does not permit the right to sue for improper care, but might hold down fees. The respondents chose Candidate A with 71% versus Candidate B with 20% (Public opinion on line, Question 075, 1998).

Although the public who was polled in the above opinion surveys seem to be overwhelmingly supportive of the Patient's bill of rights when it is dealt with as a single issue, the issue takes on less importance when the public is given a ranking for importance against other issues. In a November 1999 Roper poll, participants were asked to indicate which one health issue President Clinton and the new Congress should deal with – assuming they could only deal with one issue: Protecting patient's rights – 23%; taking steps to keep Medicare sound in the future – 37%; providing health insurance for uninsured Americans – 31% and reducing smoking among young people – 6% (Public opinion on line, Question 029, 1999).

Other opinion polls were conducted in 1999, which delineated the Republican versus Democrat positions on the Patient's bill of rights. Additional issues included in the survey were allowing appeals to be heard by independent physicians and giving doctors the final say on treatment decisions. The Democrat's proposal which included the above issues in addition to applying to all 161 million Americans, received 65% approval versus 23% for the Republican proposal (Public opinion on line, Question 024, 1999). In a similar poll, the results were 56% Democrat versus 32% Republican (Public opinion on line, Question 062, 1999). When the public was asked without labeling of Republican versus Democrat proposals, 54% were in favor of adding more provisions and 35% wanted to limit federal regulation, with 11% no answer or do not know (Public opinion on line. Question 025, 1999).

While Congress debates the two competing proposals to give patients a bill of rights, many states have passed their own regulations, which apply, to about 60% of Americans. Governors continue to work hard to include those individuals covered by self-insured plans which are exempt from state oversight due to a 1974 federal law. U.S. Representative Jan Schakowsky (D-IL) has an interactive page on her website that allows her constituents to share their "HMO horror stories". Representative Schakowsky wants to know if her constituents have been denied care, forced to change doctors in the middle of treatment, lost coverage, refused access to a specialist or had to work for days to get what they deserved. Constituents can also sign a petition calling on Congress to pass the Patient's Bill of Rights (Schakowsky, FDCH Press Release, 1999).

Health Status Outcomes

A review of the literature reflects a mixed model of findings. For example, some studies have found minimal differences between the two systems (FFS versus managed care) relative to access or quality of care (Riley et al., 1994). However, other studies have found populations of patients who are elderly or chronically ill and are at the poverty line do worse in the managed care system (Ware et al., 1996). Further, some studies showed patients were denied coverage for life-threatening illnesses or the health plans whose gatekeeping practices kept patients from utilizing emergency rooms raised concerns for health providers, patients, lawyers and society (Zautcke, 1997; Young, 1997). The reader needs to be alert to the populations being studied and the caveats given by the researchers.

A 1999 study published by Sullivan addresses the paucity of meta-review analytic studies of the quality of care variable in comparing outcomes for patients in managed care organizations versus fee-for-service. His is the third study that has been done – with the others being carried out in 1994 and 1997. His meta-analysis of 44 research studies demonstrates that the quality of care variable is equal to or inferior to that in fee-for-service plans.

A review article of health plan report cards reported the methodologic differences in collecting data for such plans diminishes the usefulness for consumers, providers and attorneys of evaluating health plans (Scanlon et al., 1998). This is the first and only study that has carried out such comparisons. Of concern is that such health plan survey data may only capture best the member's perception of convenience in using the system and communication skills of the health provider.

One limitation of the managed care model is non-continuity of care of patients with health providers. A recent local example of this is the October 1999 announcement by Mutual of Omaha that they would be ending their health maintenance organization (HMO) for Medicare patients. Their rationale was based on economics – they stated the Health Care Financing Agency (HCFA) payments were no longer cost effective for them (Health Law Update, October 21, 1999). For some patients, this could mean having to again find another health provider. There is an abundance of health research literature that correlates positive patient outcomes with having the same health provider over a continuum of time. This is especially true for individuals with chronic illnesses – as would be found with older individuals.

According to news reports of July 25, 2000, more than 900,000 elderly and disabled people will be dropped by health maintenance organizations pulling out of the Medicare program. The Clinton administration's count of 933,687 represents nearly one-sixth of the 6.2 million Medicare recipients in HMOs, and underlies the turbulence and instability of the market in which Medicare beneficiaries seek private health insurance. Medicare beneficiaries will have until January 1, 2001 to either enroll in other managed care plans, assuming they can find one in their area, or return to Medicare's original fee-for-service program. Many of these older Americans will lose the benefits provided by HMOs, which includes low-cost coverage of prescription drugs. The cut in Medicare enrollees is an indication that many in the HMO industry no longer believe they can make a profit in the Medicare business. In fact, HMOs are hoping that consumer complaints about the disruption of care will persuade Congress to boost payments for Medicare HMOs. Both Congress and policy analysts expressed disappointment that HMOs could not restrain the growth in spending for Medicare.

Research has been done on the difference between for-profit and not-for-profit HMOs. Health providers need to know the shift in membership that has occurred in the past 15 years in these types of delivery systems. For example, the percentage of individuals who have enrolled in for-profit plans has increased from 26% in 1985 to 62% in 1998 (Curtin, September, 1999). A recent study reported these differences between these two types of HMOs – the for-profit HMOs "gave fewer immunizations, fewer mammograms, fewer Pap smear tests, fewer eye exams for diabetics, and filled fewer prescription drug orders for high blood pressure for heart attack" patients as compared to the not-for-profit HMOs (Curtin, September, 1999, p. 1). The researchers, in making the above comparisons, were using the 14 quality indicators, which are used by the National Committee for Quality Assurance, a private agency that accredits health plans. However, additional concerns are noted by the following: (a) consumers paid about the same amount in both plans; (b) the for-profit HMOs spent 48% more on administrative support that the other type of plan; (c) the five largest for-profit HMOs reported 1.5 billion dollars in profit; and (d) the highest paid HMO executives made 6.2 million dollars yearly with additional stock options averaging 13.5 million dollars yearly. A concern for consumers is that now that this study has been published, for-profit HMOs are no longer releasing such data outcomes. In 1997, 41 HMOs would not release data; by 1998 this number had risen to 155. Since they are private entities, they are not required by law to disclose such data. The last section of this paper will address some of the justice issues raised by the managed care model.

Another study (Dresser et al., 1997) reinforces the above concern with having reliable outcome indicators for HMOs. The conclusion of this research team is that HMOs can provide more reliable data to health providers, employers and consumers if they use a hybrid method of reporting their outcomes, that is, data based on both chart review and automated claim data.

A different study reviewed 16 studies, which compared HMOs to fee-for-service systems. The outcomes of this study were that the quality of care of HMOs were equal to or better than care provided in the fee-for-service system. In addition, HMO members received more preventive care as demonstrated by the number of physical examinations, immunizations and breast examinations. This was done at no additional cost (Curtin, September, 1999). Thus, this meta-review study differs from the one above in the kinds of delivery systems that were compared, that is, this studied FFS versus HMO whereas the earlier study cited by Curtain compared for-profit HMOs to non-profit HMOs.

Kerr, Hayes, Mitchinson, Lee and Siu (1999) studied the relationship of gatekeeping and utilization review on patient satisfaction. Health plans that limit direct access to specialists, and especially to those specific specialists whom a patient desires, will result in decreased patient satisfaction. Further, the patient is more likely to disenroll and not recommend the plan to others. In recent years, many HMOs have responded to the findings, which are supported in this study, and have given Americans more choices in this area. HMOs have evolved and have recognized Americans' preference for choice in their physicians. At the minimum, this study demonstrates this American trait.

A study by Chin and Harrigill (1999) focused on timeliness of care for patients needing gynecologic surgical treatment and compared patients in managed care and fee-for-service systems. They found that patients in the former system experienced a delay for benign diseases but no delay for oncologic diseases. This study was partially prompted because of research which demonstrates patients with colorectal cancer who belong to a HMO have had delayed surgical treatment (Francis, 1984).

On the other hand, patients in Medicare managed care systems fared better than patients in fee-for-service systems in Washington state for the rate of influenza immunizations (Ballard et al., 1997). The seniors in the HMO plan had a higher immunization rate.

Bruno and Gilbert report that for Medi-Cal clients in California, clients are better off in a Medi-Cal managed care system than in a Medi-Cal fee-for-service system (1998). Our critique of this study indicated that all of the variables cited are process versus outcome indicators. The researchers list the following positive process indicators which are a part of the managed care system but are not required or do not exist for the fee-for-service system: (a) physician credentialing; (b) physician site reviews; (c) hospital and facility site review; (d) access to care; (e) pharmacy benefit management system; (f) member medical grievance system; (g) utilization management/speciality referral oversite; (h) case management program; (i) health education; (j) prevention programs; (k) quality of care studies; (l) encounter data reporting; (m) cultural and linguistic requirements; and (n) regulatory oversight. They do add the caveat that there has not been enough time to compare actual health outcomes of members of the two systems. Because of the strong emphasis in the health care system now on outcomes and evidence-based practice, the importance of this study, which only uses process evaluation indicators, has to be scrutinized.

A population of women studied in 1994 who belonged to an HMO demonstrated these results – the women in an HMO were more satisfied with out-of-pocket costs and the range of services they received (Wyn et al., 1997). However, they were less satisfied with the choice of provider and access to care. They were being compared to other women in employer and union-sponsored plans.

Managed care has also created barriers to women's health. In a first-ever study of how well managed care is serving the needs of low-income women, the Center for Women's Policy Studies reviewed contracts between managed care companies and 10 states and the District of Columbia, and conducted an in-depth analysis of the New Jersey contract. The study found that managed care increased the fragmentation of low-income women's health care by forcing enrollees to go outside the plan for many critical Medicaid-covered services, such as family planning, substance abuse treatment and STD/HIV treatment. In addition, the enrollees do not often know that providers serving people with Medicaid exist outside the plan. Three weaknesses were identified in Medicaid managed care plans: (a) a failure to provide a comprehensive package or Medicaid-covered services or to provide referrals outside the plan; (b) an absence of systems for consumer accountability (including advocates to help patients pursue grievances or appeals of coverage decisions); and (c) a failure to guarantee freedom of provider-patient communications and to protect the confidentiality of medical records (U.S. Newswire, 1998).

Research – Physicians and Managed Care

The managed care model has created much professional turmoil for the medical profession. For the first time in their professional history, physicians are losing autonomy as professionals, are becoming employees, and are having their medical practice become bureaucratized. While the fee-for-service delivery system had its own set of legal and ethical challenges, this section of this article will articulate the legal, professional and ethical challenges physicians now have to face in the managed care model. In one perspective, the challenges are the mirror opposite of the previous challenges. In the former system, physicians had incentives to "do more" for the patient; in the current system, the physician has incentives to "do less" for the patient. Dr. E. C. Rich has identified the three main conflicts for physicians – lack of autonomy, decreased time and the incentive to "do less" for the patient (Seminar presentation, November, 1999). The first conflict, lack of autonomy, is shown by the trend from employment as a solo practitioner or as working in a small partnership to now being increasingly employed in a large group practice. With this change in employment status has come less professional control of one's practice. The second conflict, lack of time, is demonstrated by the increased workload, the demands of the HMO re: numbers of patients expected to be seen per hour, the time and hassle factors necessary to obtain permission for patients to receive treatment, and so forth. These demands on time may lead to less advocacy by the physician for the patient. Finally, the third factor, the incentive to "do less" for the patient results from the specifics of the HMO contracts, the employment group practice incentives, and risk sharing programs. The American Medical Association is concerned about the ethical challenges which the managed care model is raising (Rich, Seminar presentation, November, 1999). All of the above concerns are and can be leading to the following: (a) physicians putting their own financial interests above the welfare of patients (which is the antithesis of the medical profession's Code of Ethics); (b) physicians losing control of their professional practice; and (c) the potential erosion of the public's confidence in physicians' competence. This latter fear is supported by the following trend noted in a Gallop Poll. In 1983 40% of Americans thought the medical profession was an admirable profession. By 1996 that number had eroded to 14% (Rich, Seminar presentation, November, 1999). In addressing all of these concerns, Dr. Rich advocated these five strategies to problem solve: (a) use evidence-based practice as one's basic practice pattern; (b) improve one's medical practice through research; (c) be an actor who is redesigning financial incentives; (d) reward service and quality; and (e) practice a community orientation, that is, be culturally competent, provide community service and be involved in public education and advocacy.

An example of the pressures physicians are under – the tension between following aspects of the professional medical ethics Code and risking economic loss – is demonstrated by McNamara et al.'s study (1998). They note the capitation system as being an increasing model of payment for physicians – one third of physicians had at least one HMO contract. Further, this payment model is extending to specialist physicians as well as primary care providers. They discuss the concerns noted in the above paragraph – "capitation provides a disincentive to select individuals with chronic illnesses who have high expected costs" (McNamara et al., 1998, p. 1178). To facilitate cardiologists' non-economic loss, they studied the best way to anticipate costs of patients with coronary artery disease (CAD). At present, capitation rates are set based on a Medicare data set and the variables of age and gender are used. Their study found the best predictors of cost were not age and gender; rather, the measures of severity and comorbidity were more predictive. Implications of this study for cardiologists are to: (a) know of such research; (b) negotiate contracts based on this knowledge (negotiate for higher re-imbursement if the patient population has such measures of severity and comorbidity); and (c) use re-insurance to prevent financial disaster.

Nolan has written about another physician specialty, internal medicine, and raises the question – "does the traditional relationship between internist and patient still have a role?" (1998, p. 857). He states the managed care model has significantly changed the role of this physician specialist because of the expansion of the role of the family practice physician and the introduction of the "hospitalist", a specialist who takes responsibility for patients in the hospital area. Thus, the role of the internist has eroded. He distinguishes family practice from internal medicine. He deplores "Managed care, with its emphasis on a high volume of 10 to 15 minute encounters, [which] works against the very core of what makes internal medicine the productive and intellectually stimulating discipline it has been over the past century" (Nolan, 1998, p. 859). He advocates the internist as being more prepared to care for patients with complex chronic illnesses and better skilled to implement certain select technologies with which they have had more practice. He also makes the economic argument – since 80% of the health care costs in the United States are related to individuals with chronic diseases, it is economically wiser to care for those individuals with internists who are the most prepared. He critiqued the increasing use of "hospitalist" physicians because the revered continuous physician/patient relationship is no longer present and the total knowledge of the patient and his condition may not be as known to the "hospitalist" physician as it would have been to the internist. While one may critique Nolan in that he is struggling to hold onto a dinosaur that is no more, the article does reflect the turmoil for physicians in this changing health system. A contrasting perspective on the value of the "hospitalist" or "intensivist", the language used in some literature, is the finding that there is decreased mortality for a sub-set of patients (those with abdominal aortic bypass surgery) when such an "intensivist" does daily rounds (Curtin, August, 1999).

Also, in contrast to Nolan's essay is a research study of internal medicine residents conducted by Nelson, Matthews, Patrizio and Cooney (1998). Their research found 21% of these residents believed the managed care model was the best system and 31% said they would be satisfied working in such a system.

Hillman, a physician, writes more positively of managed care (1998). He notes the negative perverse incentives for physicians in the fee-for-service system. He challenges health providers to problem solve in two areas: (a) sanction management situations that place profit over patients; and (b) implement quality assurance mechanisms. He praises the new tools that are being used and are an outgrowth of the managed care model – outcomes evaluation, quality assurance, disease prevention, disease management, meta analysis, decision analysis, cost-effectiveness analysis, and so forth.

Across the United States, physicians are complaining about the effects of managed care. Nine in ten physicians, approximately 540,000, have at least one managed care contract. A survey of 900 doctors aged 40 and younger found seven in ten were dissatisfied with their relationship with managed care and just more than half identified "denial of care" by health plans as their greatest concern (Omaha World-Herald, 3/14/99). Some physicians are dropping HMO contracts, forming professional alliances and unions, filing lawsuits and disability claims, and taking early retirement. The complaints about managed care include robbing physicians of their autonomy, quality, income, time, prestige and self-respect.

Research – Nurses and Managed Care

During this decade one of the spin-offs of the managed care model has been a re-structuring of the personnel employment pattern in hospitals and other health organizations. One such result has been the 'down-sizing,' and 'right-sizing' of nurses and the increased replacement of nurses with unlicensed assistive personnel. With such loss of employment has come both research and legislation. A recent study reported the single most important predictor of positive outcome indicators for patients in acute care institutions was the percentage of nurses (RN) on the nursing staff (Moore, 1999). Patients, consumers, nurses and physicians can utilize this knowledge in advocating for legislation and/or analyzing fact patterns in litigation. In the Fall of 1999 legislation was passed in California which mandates certain nurse/patient ratios. This is the first of its kind in the country and reflects the concerns discussed in this paragraph.

Research surveys report nurses' concerns with the managed care model. A recent survey by the Kaiser Family Foundation has received much publicity. In that survey of nurses and physicians, the following results were noted. Eighty percent of nurses stated that managed care has decreased the quality of care for ill patients (Stewart, 1999). Fifty percent of nurses reported that decisions of health plans have resulted in the decline of the health status of patients. Of these latter nurses, two-thirds of the nurses stated such adverse decisions occurred weekly or monthly. In this survey, 69% of the nurses said inadequate staffing was their greatest concern. In contrast, 58% of physicians said the increased administrative time factor was their biggest concern (Stewart, 1999). In addition, 90% of physicians reported that they had patients who had been denied coverage, which resulted in adverse outcomes (Kaiser Survey, 1999). Forty eight percent of nurses cared for patients who had been denied coverage. Health providers' advocacy responses to the above were noted by: (a) two-thirds of the physicians reported they sometimes or often intervened to advocate for their patients; and (b) 50% of nurses and physicians said they exaggerated a patient's severity of condition in order to obtain coverage for the patient (Kaiser Survey, 1999; Health Lawyers News, 1999). Ninety percent of both physicians and nurses stated they spend far more time doing paper work and 80% of both health providers said they have less time to spend with the patient. When there is denial of coverage, the categories were as follows: prescription drugs, 61%; diagnostic tests, 42%; overnight hospital stays, 31% and referrals to specialists, 29%.

A second study, that of 900 nurses by the New York State Nurses Association (NYSNA), found that 50% of those nurses reported they could not provide the type of adequate nursing care their patients needed (Stewart, 1999). Such concern is noted at a time when 80% of hospital nurses speak of the increased acuity of patients in hospital settings.

Both of the above studies support a 1996 study done by the American Journal of Nursing (AJN) in which 90% of nurses voiced concern about the decreased safety and quality of care for patients because of downsizing of Rns and the employment of unlicensed personnel (Stewart, 1999). The AJN study was repeated in 1997 and there was an increase in nurses who rated their hospitals as "poor" or "very poor" in such indicators as patient/family complaints, patient injuries, medication errors, pressure ulcers, skin breakdown, iatrogenic illnesses, and so forth. (Stewart, 1999).

A study by Buerhaus and Staiger (1999) mapped the employment rate of nurses and the geographic market penetration rate of HMOs. They found the employment rate of nurses and their earnings growth slowed between 1994 and 1997 and this change occurred first in the high HMO penetration markets.

The stance taken by several leaders and organizations within the nursing profession has been mixed about the managed care model. Bev Malone, immediate past president of the American Nurses Association, sums up this duel message best. She notes the positives of the model – continuity of care [which can be argued and critiqued], case management, data collection, research, increased emphasis on prevention, outcomes studies and evidence-based practice. However, she notes that some of the short-term cost-effective gains have been realized at the cost of long term patient outcomes. She deplores the negative patient outcomes that have occurred because of profit (Stewart, 1999)

A study of 1000 nurses in Minneapolis/St. Paul hospitals, an urban area with a high penetration of managed care, identified patient care concerns when unlicensed assistive personnel (UAP) are used (Preuss, Sharing Care, 1998). A major concern was that certain tasks were being delegated to UAPs – but the unexpected result is that much crucial information was being lost to the professional nurse. A variety of recommendations come from this study of how to best meet patient needs by nurses in this changing environment.

FUTURE OF MANAGED CARE

There is concern with the passage of the Patient's Bill of Rights. Some want it to be a stronger document. Republicans argue that if it had passed as proposed it would have increased HMO premiums by 15% to 30% which would have increased the cost of health care by $100 billion dollars resulting in two million Americans losing their health insurance and 240,000 individuals losing their jobs. Another perspective is the estimate done by both Coopers & Lybrand and the Kaiser Family Foundation. They estimated the increase would be 0.61% to 0.77% or one billion dollars – or $3 per member per month. This debate continues in President Bush's administration. When the issue is framed to fuel the fears of those who already have health insurance and that they may lose it, plus lose their job, such individuals may not have the interest to further the passage of a stronger legislative bill (Curtin, September, 1999).

Curtin, in analyzing two articles from a recent Wall Street Journal, thinks that the Supreme Court may hold against HMOs for profiteering at the expense of people's lives (November, 1999). She compares a current HMO case before the Supreme Court with the recent Supreme Court decision that levied 1.2 billion dollars in punitive damages against GMC. Judge E. G. Williams wrote, "The court finds clear and convincing evidence demonstrating that defendant's fuel tank was placed behind the axle on automobiles of the make and model here in order to maximize profits – to the disregard of public safety" (Curtin, November, 1999, p. 1). She makes the analogy – if the Supreme Court is balancing the value of life compared to the $8.59 per vehicle that GMC could have spent to prevent this problem, what will the Supreme Court hold in the case before the Court of a patient who was endangered because of a HMO's denial of treatment? Will the Supreme Court use the same public endangerment rationale they used in the GMC case? She thinks and hopes they might.

Some writers voice concern about the future of the managed care model into the 21st century because of rising premium costs, the implications of the proposed Patient's Rights bill, and the large penalties which are being assessed by courts against HMOs because of denial of health care (Stahl, November, 1999). Illinois is one state that has enacted laws, which allow patients to sue HMOs if their physicians fail to provide services. Thus far, the largest award in this regard has been 120 million dollars in the case of Goodrich v. Aetna U.S. Healthcare (Stahl, November, 1999). To prevent the demise of HMOs, consultants recommend these three strategies: (a) control increasing costs by creating centers of excellence, that is, for diabetes, and so forth; (b) use "best practices" models of care – thereby decreasing morbidity and increasing positive outcomes; and (c) increase research on the effectiveness of care management.

Friedman advocates the future of managed care depends on the moral performance of current health plans and of pushing the bad plans out of the system. She describes one extreme negative example – the HMO that obtains a contract for Medicaid patients, assigns the population to a primary care provider who does not intend to care for them and takes his/her 30% cut (1999).

In 20 years, what will the historical analytical articles say of the managed care phenomena? What will be written of the health providers? Were they active enough actors in their professional roles to advocate for patients? Was it the tool of the legislatures (state and federal) that enacted legislation to change policy? Was it the tool of the courts and litigation that changed policy?

Pinkerton is one who advocates for the continuation of HMOs because they are better than a "federal takeover" (August 16, 1999). He dismisses the ideas of "four leading leftists", Drs. David Himmelstein, Steffie Woolhandler, and Ira Hellander of the Physicians for a National Health Program and Sidney Wolfe of the Public Citizen Health Research Group, "a Naderiste organization" (August 16, 1999, p. x). He critiques the money being spent on litigation in the health care field and the increasing regulation of HMOs: (a) $20 billion yearly on malpractice; (b) $60 billion yearly spent on defensive medicine; and (c) the potential passage of the Patient's Bill of Rights which will make HMOs most costly to operate. He posits that these kinds of expenses will increase the costs of HMOs and make health insurance less available and that we will see a larger percentage of uninsured Americans.

The interplay of all aspects of the health, legal and economic systems of this country must be noted. For example, articles published about the legal response to problems with the managed care system also specifically note what happens to the stocks and profits of HMOs when court rulings are handed down and/or when lawsuits are initiated. For example, in early October 1999, stock prices in HMO companies that were involved with lawsuits decreased 21%, 18%, and 20% (Leibold, 1999).

The problems raised by the managed care model may eventually take the American public back to some questions raised during the early 1990s. What model do we want? Do we want health insurance for all? Do we want some basic level of care for everyone? If so, do we then want additional health care based on one's ability to pay? Leibold (1999) notes that when the private sector managed care model "won the day" in the early 1990s, HMOs were to decrease costs and improve quality. Research is still pending on quality improvement. In the meantime, quality of and access to health care has not improved for many – and, especially to people with voice, that is, educated, middle-class, employed individuals. One research study demonstrates there is less access to health care for low-income uninsured persons in those areas that have a higher penetration of Medicaid managed care (Cunningham, 1999). The rationale for this finding is the emphasis on cost-savings prevents cross-subsidization of care to medically indigent. This research, the first study of its kind, defuses some of Pinkerton's arguments noted above. One could posit the argument that HMOs have been positive for selected actors – but, not for many other individuals, that is, the ill individuals denied treatment, the individuals who have received less than quality care, and, as Cunningham's research demonstrates – the low income uninsured populations living in such geographic areas as noted above. Leibold notes that legislation will continue being passed at the state and federal level to address specific issues with managed care and lawsuits will also continue as one tool to address concerns. However, the larger political will question of the American public will remain – is there a national consensus – is health care a right?

The October 1999 issue of the Journal of Health Politics, Policy and Law is devoted to the phenomena of the managed care model. Stone is one of several writers (1999) who addresses a backlash against this model and she speaks of a Second Great Transformation. "The essential message of all the horror stories told by patients is the anguish of abandonment. The howl of doctors, nurses, and other caregivers is moral revulsion at the callousness they are forced to enact" (p. 1217).

SUMMARY

This paper has addressed several aspects of managed care – the predominant model of health care delivery in this country. The final answer is still pending on its outcome value. However, the problems it has created have resulted in much state and federal regulatory legislation and much litigation – and, this response has been described. Many challenges are ahead. Stone's analysis may be best – there is a backlash against the managed care model and this backlash is a revolt against the moral and social – not material – losses that are occurring in our society (1999, p. 1217).

REFERENCES

Ballard, J. E., Liu, J., Uberuagua, D., Mustin, H. D., & Sugarman, J. R. (1997). Assessing influenza immunization rates in Medicare managed care plans: A comparison of three methods. *Joint Commission Journal on Quality Improvement, 23*(8), 434–442.

Bruno, R., & Gilbert, B. P. (1998). In California, Medi-Cal managed care is superior to Medi-Cal fee-for-service. *Managed Care Quarterly, 6*(4), 7–14.

Center for Women's Policy Studies shows how managed care creates barriers to women's health. (1998, February 20). U.S. Newswire. Retrieved January 3, 2000 from Electric Library on World Wide Web: http://www.elibrary.com

Chin, S., & Harrigill, K. M. (1999). Delay in gynecologic surgical treatment: A comparison of patients in managed care and fee-for-service plans. *Obstetrics & Gynecology, 93*(6), 922–927.

Cunningham, P. (1999). Pressures on safety net access: The level of managed care penetration and uninsurance rate in a community. *Health Services Research, 34*(1), 255–270.

Curtin, L. (1999, August). The day is coming. *CurtinCalls, 1*(9), 1–2.

Curtin, L. (1999, November). A cost benefit analysis of someone else's life. *CurtinCalls, 1*(12), 1–2.

Curtin, L. (1999, September). Rebels without a clue. *CurtinCalls, 1*(10), 1–2.

Donaldson, M. S. (1998). Accountability for quality in managed care. *Journal on Quality Improvement, 24*(12), 711–725.

Dresser, M., Feingold, L., Rosenkranz, S. L., & Coltin, K. (1997). Clinical quality measurement. *Medical Care, 35*(6), 539–552.

Francis, A. M., Polissar, L., & Lorenz, A. B. (1984). Care of patients with colorectal cancer: A comparison of a health maintenance organization and fee-for-service practice. *Medical Care, 22*(5), 418–429.

Friedman, E. (1999). Truth or consequences. *Health Forum Journal, 42*(1), 8. Retrieved January 3, 2000 from EBSCOhost database on World Wide Web: http://webnfl.epnet.com

Health Law Update, 7(20), 3. (1999, October 31).

Health Lawyers News, 20.

Hillman, A. (1998). Commentary: Burden of proof in an era of outcomes research and managed care. *Health Services Research, 33*(1), 75–78.

Kaiser Survey. (1999, August 20). *Capital Update, 17*, 7.

Kerr, E. A., Hays, R. D., Mitchinson, A., Lee, M., & Siu, A. (1999). The influence of gatekeeping and utilization review on patient satisfaction. *Journal of General Internal Medicine, 14*(5), 287–296.

Leibold, P. (1999). A year of living dangerously for the managed care industry. *Health Lawyers' News, 3*(11), 5.

McNamara, R. L., Powe, N. R. Shaffer, T., Thiemann, D., Weller, W., & Anderson, G. (1998). Capitation for cardiologists: Accepting risk for coronary artery disease under managed care. *American Journal of Cardiology, 82*(10), 1178–1182.

Nelson, H. D., Matthews, A. M., Patrizio, G. R., & Cooney, T. G. (1998). Managed care, attitudes and career choices of internal medicine residents. *Journal of General Internal Medicine, 13*(1), 39–42.

Nolan, J. P. (1998). Internal medicine in the current health care environment: A need for reaffirmation. *Annals of Internal Medicine, 128*(10), 857–862.

Omaha World Herald, March 14, 1999.

Pinkerton, J. (1999, August 16). *HMOs better than federal takeover*. Omaha World Herald.

Preuss, G. (1998). *Committing to care*. Economic Policy Institute. Washington, D.C.

Preuss, G. (1998). *Sharing care*. Economic Policy Institute. Washington, D.C.

Public opinion on line. Question number 009. (1998, August 14–September 20). University of Connecticut, Roper Center. Retrieved January 3, 2000, from Lexis-Nexis Academic Universe database on World Wide Web: http://web.lexis-nexis.com/universe

Public opinion on line. Question number 024. (1998, July 13–14). University of Connecticut, Roper Center. Retrieved January 3, 2000, from Lexis-Nexis Academic Universe database on World Wide Web: http://web.lexis-nexis.com/universe

Public opinion on line. Question number 025. (1999, July 13–14). University of Connecticut, Roper Center. Retrieved January 3, 2000, from Lexis-Nexis Academic Universe database on World Wide Web: http://web.lexis-nexis.com/universe

Public opinion on line. Question number 029. (1999, November 4–December 6). University of Connecticut, Roper Center. Retrieved January 3, 2000, from Lexis-Nexis Academic Universe database on World Wide Web: http://web.lexis-nexis.com/universe

Public opinion on line. Question number 036. (1998, July 25–27). University of Connecticut, Roper Center. Retrieved January 3, 2000, from Lexis-Nexis Academic Universe database on World Wide Web: http://web.lexis-nexis.com/universe

Public opinion on line. Question number 039. (1998, July 25–27). University of Connecticut, Roper Center. Retrieved January 3, 2000, from Lexis-Nexis Academic Universe database on World Wide Web: http://web.lexis-nexis.com/universe

Public opinion on line. Question number 040. (1998, July 25–27). University of Connecticut, Roper Center. Retrieved January 3, 2000, from Lexis-Nexis Academic Universe database on World Wide Web: http://web.lexis-nexis.com/universe

Public opinion on line. Question number 062. (1999, July 24–26). University of Connecticut, Roper Center. Retrieved January 3, 2000, from Lexis-Nexis Academic Universe database on World Wide Web: http://web.lexis-nexis.com/universe

Public opinion on line. Question number 075. (1998, September 10–13). University of Connecticut, Roper Center. Retrieved January 3, 2000, from Lexis-Nexis Academic Universe database on World Wide Web: http://web.lexis-nexis.com/universe

Public opinion on line. Question number 123. (1998, June18–21). University of Connecticut, Roper Center. Retrieved January 3, 2000, from Lexis-Nexis Academic Universe database on World Wide Web: http://web.lexis-nexis.com/universe

Riggs, J. E. (1996). Managed care and economic dynamics. *Archives of Neurology, 53*(9), 856–858.

Riley, G. F., Potosky, A. L., Lubitz, J. D., & Brown, M. L. (1994). Stage of cancer at diagnosis for Medicare HMO and fee-for-service enrollees. *American Journal of Public Health, 84*(10), 1598–1604.

Rising number of doctors cut links to HMOs. (1999, March 14). *Omaha World Herald*, p. 15A.

Scanlon, D. P., Chernew, M., Sheffler, S., & Fendrick, A. M. (1998). Health plan report cares: Exploring differences in plan ratings. *Journal of Quality Improvement, 24*(1), 5–20.

Schakowsky, J. (1999, June 23). Schakowsky unveils new interactive page on web site to combat HMO abuses and build support for the Patients' Bill of Rights. FDCH Press Releases. Retrieved January 3, 2000, from EBSCOhost database on World Wide Web: http://webnfl.epnet.com

Stewart, M. (1999). Survey highlights nurses' concern about health care. *The American Nurse, 31*(5), 3.

Sullivan, K. (1999). Managed care plan performance since 1980: Another look at 2 literature reviews. *American Journal of Public Health, 89*(7), 1003–1008.

Tumulty, K., & Dickerson, J. (1998). Let's play doctor. *Time, 152*(2), 28–35. Retrieved January 3, 2000, from EBSCOhost database on the World Wide Web: http://webnfl.epnet.com

Ware, J. E., Bayliss, M. S., Rogers, W. H., Kosinki, M., & Tarlov, A. R. (1996). Differences in 4-year health outcomes for elderly and poor, chronically ill patients treated in HMO and fee-for-service systems. *Journal of the American Medical Association, 276*(13), 1039–1047.

Wyn, R., Collins, K. S., & Brown, E. R. (1997). Women and managed care: Satisfaction with provider choice, access to care, plan costs and coverage. *Journal of American Medical Women's Association, 52*(2), 60–64.

Young, G. P. (1997). Adverse outcomes of managed care gatekeeping. *Academic Emergency Medicine, 4*(12), 1129–1136.

Zautcke, J. L., Fraker, L. D., Hart, R. G., & Stevens, J. S. (1997). Denial of emergency department authorization of potentially high-risk patients by managed care. *Journal of Emergency Medicine, 15*(5), 605–609.

ON BEING A PROACTIVE HEALTH CARE CONSUMER: MAKING AN "UNRESPONSIVE" SYSTEM WORK FOR YOU

Eva Kahana and Boaz Kahana

ABSTRACT

This chapter focuses on the proactive adaptations that patients can under-take to successfully navigate today's increasingly complex health care system. Our report urges a paradigm shift in patient and family expecta-tions, moving from expectations of "patiently" awaiting good care to taking responsibility for finding and getting optimal health care services. We consider the role that key informal health care partners (caregivers) can play in facilitating these proactive adaptations. Based on extensive partic-ipant observations by the authors during recent illness episodes, we present a framework for planning and implementing proactive adaptations which can improve the health care that consumers receive.

Changing Consumers and Changing Technology in Health Care and Health Care Delivery,
Volume 19, pages 21–44.
ISBN: 0-7623-0808-7

CHALLENGES FACING TODAY'S HEALTH CARE CONSUMER

The ideals of patient-centered care have been increasingly advocated in the health services research literature (Gerteis et al., 1996; Stewart et al., 1996). In our own prior work we have outlined principles for providing patient responsive care which are likely to improve the health care experiences of consumers (Kahana et al., 1999a, b). Medical sociologists have also argued that trust in physicians and the health care system is essential to patient well-being and recovery (Kao, 1998; Mechanic, 1998). Yet, it is also recognized that the current U.S. health care system faces many challenges before patient centered care becomes the norm for care delivery, and patient trust can be deserved by the system (Macklin, 1993). In fact, there has been considerable scientific and public attention directed at the prevalence of unresponsive care, which often ignores even basic adages of medical care to "do no harm" (Rosenthal, 1999; Liang & Cullen, 1999). The many failures of self regulation in medical care for averting medical mistakes have been well documented (Rosenthal, 1999). Trust and confidence in the health care system have also been diminished with the advent of managed care, through increased discontinuity of health care, which undermines foundations of the doctor-patient relationship (Flocke et al., 1997; Kahana et al., 1999, 1999b).

Even when physicians uphold ethical ideals of advocacy to serve the best interests of their patients, professional autonomy of physicians has been restricted by real-life circumstances. Social changes affecting health care delivery have resulted in physicians being able to spend only limited time with patients and having limited opportunities to advocate for patients (Macklin, 1993). Finally, we must also recognize the many uncertainties and human limitations involved in medical care delivery, even under the best of circumstances, and in the most supportive health care climates. As Groopman puts it in his perceptive book *Second Opinions* (2000), "The bond of mutual vigilance between doctor and patient is forged through the melding of knowledge and intuition" (p. 230).

THE BROADER CONTEXT OF UNRESPONSIVE CARE

Most people can resonate to observations in the media about the increasing unwieldiness of the health care system (Goldberg, 2000). Below we summarize some of the widely recognized challenges posed by the health care climate within which consumers receive their health care today (Groopman, 2000; Lerner & Schwartz, 2000; Macklin, 1993; Pescosolido et al., 1997).

- Bidding for contracts with physicians by managed care insurers has led to disrupting long-term relationships between patients and physicians (Flocke et al., 1997).
- Physicians' autonomy in treating patients has been curtailed (Hafferty & Light, 1995).
- Managed care has led to hospitals discharging patients "quicker and sicker" (Pescosolido et al., 1997).
- Patients' options for seeking out specialists and state-of-the-art health care have been limited with the shrinking of fee-for-service medical care (Macklin, 1993).
- Major advances in technology have led to increasingly complex procedures for diagnosing and treating illness (Moloney & Paul, 1996).
- The information explosion has limited familiarity of even good physicians with fast-paced research advances in diagnosing and treating illness (Wholey et al., 2000).
- The development of large hospital conglomerates has made communication among physicians treating a patient increasingly difficult (Macklin, 1993).
- Intensive and ongoing contact between patients and their physicians has diminished and been replaced by an assembly line system of medical encounters, where the patient comes into contact with a multitude of clerks, schedulers, and ancillary personnel who are often geographically scattered (Groopman, 2000).
- Standard practices in most hospitals reinforce expectations of dependency and passivity among patients (e.g. patients are required to use wheelchairs to go for medical tests even when they can walk without difficulty) (Gerteis, 1996).

Anticipating these developments, the American Hospital Association published the landmark "A Patient's Bill of Rights" in 1973. This bill of rights enunciates the right to considerate and respectful care, as well as the right of access to all records pertaining to one's care and complete current information about diagnosis, treatment, and prognosis. Yet, experiences of patients who are hospitalized routinely reveal large gaps between these ideals and prevailing practices (Macklin, 1993). There have been some other notable advances in protecting patient rights in recent years. Legislation has been employed to protect the rights of patients by ensuring the right to appeal, ombudsmen in hospitals, informed consent procedures, and legal remedies for malpractice (Allshouse, 1996). Nevertheless, these formal solutions are at odds with the implicit trust between patient and physicians (Mechanic, 1998). Most patients have enough common sense to know that formal remedies come into play only when the battle for good care has already been lost. Just as seeking formal remedies has little practical value for averting bad outcomes during an illness episode, drastic solutions recommended by critics of the health care system tend to be unfeasible or unpalatable to most people as they seek health care. Nor are adversarial approaches likely to bring satisfying results. Patients who challenge hospital routines are labeled as uncooperative "problem" patients (Lorber, 1975). Thus, the major challenge facing health care consumers relates to finding avenues for optimizing their own health care, even as we collectively await systemic improvements in health care delivery to materialize.

In the medical sociology and health services literatures there have been two divergent avenues noted for addressing perspectives of the key stakeholder, the patient, on challenges in obtaining good medical care. One is a bottom-up approach wherein the solution is seen in patients' challenging medical authority and demanding to share in decision-making power with the physician (Haug & Lavin, 1978). While this approach, often referred to as consumerism, has merit, it is based on some misplaced expectations. This approach assumes that the major problems experienced by patients in obtaining good care relate to the physician-patient encounter. In fact, some of the most frustrating experiences of patients in receiving care involve interactions with non-physician personnel who often block access to physicians.

Another fallacy of this conflict-driven model relates to the realities of power relations in confrontations between the patient and the health care system. Typically, the patient is too vulnerable, and dependent on the goodwill of health care providers, to be able or willing to risk confrontations. Furthermore, the time-frame involved in obtaining formal conflict resolution is typically too long to benefit the patient during a given illness episode. In our discussion, we do not reject consumerist approaches, but instead suggest reorienting them. We urge movement from conflict-driven approaches based on rejecting the system to resolution-driven approaches which aim to make the system work for the patient.

The second approach to enhancing responsiveness of the health care system is a top-down method, calling for patient-centered care, based on health care provider and particularly physician sensitivity to the patient's subjective experience. In principle, patients could acquiesce, according to this view. The patient-centered approach to care is considered to be the norm, and thus patients are routinely expected to receive responsive care. This approach is generally seen in the health care provider literature as the vehicle to improved patient care (Stewart et al., 1996; Gerteis et al., 1996). We applaud the goal of top down approaches and see the benefits in patient involvement in enlisting advocacy of their physicians to maximize responsiveness of their health care. Nevertheless, there are also several inherent problems in top down approaches as well. Even when the patient's perspectives are noted, in the health services literature, it is often with a patronizing tinge. The patient's views are described as "subjective", rather than the "objective" medical view. Even as the value of considering the patient's perspective is acknowledged, such patient-centered care is often based on judgments made by professionals. Sometimes, it is the health professional's appraisal of what he or she would want under similar circumstances, which get reported as the patients' perspective (Brennan, 1995). Furthermore, we are often told that cultural goals of patients must be recognized, implying that the patients' perspectives on their illness are largely shaped by factors unrelated to the demands of illness (Moloney & Paul, 1996).

It has been noted that the value orientations of health care professionals often call for individualized treatment of each patient. Nevertheless, actions of these same health care professionals toward diverse groups of patients tend to be generalized and undifferentiated (Quint, 1972). As a result, inattention or unresponsiveness of hospital staff to their individual needs and perspectives regularly frustrates patients. These patient dilemmas constitute important paradoxes of hospitalization and examples of person-environment incongruence or misfit, which are likely to add to the stress of hospitalization for the patient (Kahana, 1980). In an effort to alert patients and their caregivers to areas of needed vigilance and advocacy, it is useful to review paradoxes of hospitalization from the patient's perspective. Examples of these paradoxes are noted below, and suggest the ways in which hospital life may pose particular stressors for patients, based on disregard for the patient's perspective and experiences, and based on a management-based, rather than patient-responsive, philosophy of care (Kahana et al., 1999b). Using a sociological framework, it is noteworthy that many of these paradoxes may be translated into norms or expectations which are unresponsive to the patient's phenomenology. In fact, patterns of communication which prevail in hospitals are likely to disempower rather than empower patients, and set up barriers to patient's advocacy on their own behalf. These paradoxes of hospital life exemplify a model wherein patients are expected to accommodate, and be responsive to, needs of the staff and of the hospital organization.

PARADOXES OF HOSPITALIZATION FOR THE PATIENT
(adopted from Kahana et al., 1999b)

(1) Expected to be compliant when stakes are highest for being assertive.
(2) Given minimal information when knowledge could be greatest source of power.

(3) Can't initiate communication with physician when most in need of his/her responsiveness.
(4) Expected to be optimistic and cheerful to visitors even when feeling down.
(5) Expected to be available to all when most needing privacy.
(6) Expected to accommodate to routines set by others when most in need of personalized attention.
(7) Deprived of customary environment and routines when most vulnerable to change.
(8) Exposed to interminable waits when needs are most pressing.
(9) Must be vigilant to surroundings amidst loss of normal cues and capacities.

In this paper we call for a unified effort by patients and their informal family caregivers to forge partnerships and linkages with potentially supportive individuals within the health care system. Further, patients and their advocates should also be willing to seek proactive solutions which involve shaping and influencing the system when necessary. Since the patient is not overtly recognized as the major stakeholder and a knowledgeable partner by an unresponsive system, this calls for vigilance, assertiveness, and advocacy in obtaining good care. Informal health care partners play a critical role in facilitating such needed proactivity, particularly since at times of serious illness and hospitalization the patient is likely to lack the physical and emotional resources needed for proactively navigating the health care system.

METHOD

This paper is based on participant observations by the co-authors during three recent serious illness episodes. During these illness episodes both the patient and the health care partner spouse took extensive notes. In addition, they completed brief structured health care encounter forms for each significant interaction with a representative of the health care system. Health care representatives included diverse personnel such as scheduling clerks, secretaries, nurses, physicians, hospital librarians. The adaptive tasks for patients as well as the strategies used to deal with these tasks were abstracted from notes and encounter forms. We first focus on principles derived from our observations and then turn to more detailed illustrations of the unresponsive patterns observed within the health care delivery system. We also recount how proactive adaptations were found to be useful in eliciting desired results. We want to acknowledge that the principles we describe have been derived from experiences of well-educated patients who have the benefits of health insurance, and that our examples of unresponsiveness in the health care system may be even greater for uneducated and poor patients who may also need professional advocates to be able to act as proactive health care consumers (Abraham, 1993).

THE PERSONAL CONTEXT OF OUR OBSERVATIONS

The co-authors are resourceful but very average health care consumers. For many uneventful years a reliance on the health care system served us just fine. We were in generally good health and could laugh at or lament small frustrations in getting health care. We did not consider how vulnerable we could become if a serious health crisis struck without warning. During the past year, just such dreaded health problems suddenly struck. These were health problems that could not have been averted by healthy lifestyles and sound self-care practices.

As a means of coping with the very dark hours we all so fear, we took detailed notes and tried to use all of our creativity and proactivity to survive these threatening illness situations. We tried to distill the many unexpected lessons we learned, and in this paper we share strategies that worked for us to obtain excellent care even against great odds. We found that even if warnings about the dangers of getting sick in America are justified (Macklin, 1993), there is still a great deal that patients and their informal partners can and must do if they want to win their battles for improved health.

While every patient faces unique challenges and has differing resources, supports and health care access, we have been impressed by many shared problems and shared solutions applicable to different illness situations. Here, we would like to share examples of successful efforts for overcoming the limitations of the system. Our long-range goals are to point to areas of needed change for the health care system. But systemic change is slow to occur, and by the time structural changes occur, it may be too late for us or our loved ones as we are experiencing illness situations. So here we sketch out a survival guide for dealing with the many illogical, unexpected and hard-to-recognize hazards that we face in getting the best care for what ails us.

Often problems posed by unresponsive care seem frustrating and insurmountable. As vulnerable patients, we feel manipulated and powerless at a time when we need help the most. Our suggestions, based on our own experiences, point to ways of getting around deficiencies of the system and try to make the system work for us.

The focus of our discussion in this paper is on proactive patienthood, or efforts to deal with, get around, or master an unresponsive health care system.

We recognize, of course, that there are many instances of positive care and responsive health care delivery. We specifically note that patient initiatives toward building a strong and enduring relationship with key health care providers can facilitate responsiveness from the system. Furthermore, we must also recognize that responsive, as well as excellent care often exists at institutions which specialize in specific health problems or those that have created organizational climates enhancing responsive care. Proactive adaptations often involve an active search for centers of excellence for given health problems or finding a more responsive health care provider or health care environment. Thus, we do not imply that responsive care does not exist, but rather suggest that it may be elusive, hard to find, or that its attainment may require active maneuvers on the part of the patient and the family health care partners.

The following short outline summarizes the principles of proactive consumerism in health care. Each of these principles will be further discussed throughout the remainder of this paper, with illustrations from our own health care experiences.

PRINCIPLES OF PROACTIVE PATIENTHOOD AND CONSUMERISM IN HEALTH CARE

I. Preventive Adaptations

A. Developing relationships with health care providers which enhance responsive care.
B. Recognizing the adaptive tasks of the patient:
   1. Gaining access to health care services.
   2. Gaining access to information needed for proactive involvement in health care.
   3. Obtaining financing for health care.
   4. Enhancing comfort and quality of life.
C. Examining assumptions about the responsiveness of health care system.
D. Using proactive approaches to deal with unresponsiveness in the health care system.

II. Corrective Adaptations

A. Linking up with responsive people behind the unresponsive system.
B. Gathering and using information about options proactively.
C. Communicating effectively with representatives of the system.

FACILITATING ONE'S OWN CARE

I. Preventive Adaptations

A. Developing Relationships with Health Care Providers to Facilitate Responsive Care.

While this paper is concerned primarily with proactive strategies for identifying and coping with unresponsive health care, it is important to recognize that patients can take preventive steps toward enhancing responsiveness of their care in the future. Perhaps a key step in this direction is that of developing and fostering a long-term personal relationship with primary care physicians, specialists and ancillary health care personnel such as receptionists and nurses in offices where one regularly receives outpatient care. The maintenance of long-term, personal, human relationships with key health care providers can go a very long way towards insuring responsive care when problem situations arise. A good example of the value of such relationships is reflected where the patient finds himself or herself at an emergency room. A primary care doctor or specialist who has had a long-term relationship with the patient is far more likely than a new or little-known doctor to follow up on emergency care, even as the crisis unfolds. Such a long-term relationship may prompt a personal visit to the emergency room from the physician when he or she is notified that the patient is being seen there.

Proactive efforts must often be undertaken by patients to succeed in maintaining enduring bonds and long-term associations with their primary care physicians. Renegotiating of contracts between insurance companies and health care providers has increasingly resulted in forced disruption of doctor-patient relationships, with patients required to find new physicians, since insurance companies only reimburse for services of "preferred providers" (Flocke et al., 1997).

In order to continue to be seen by long-term physicians, patients must use ingenuity and at times be willing to shoulder a greater proportion of the health care costs and/or put up with inconveniences of filing reimbursement forms to gain even partial reimbursement for services (Kahana, Stange, Meehan & Raff, 1997). Unwillingness to accept "rationing by inconvenience" and agreeing to switch doctors often results in receiving care from unfamiliar doctors and less assistance from doctors in enhancing responsiveness of the system. When patients are unwilling or, more likely, unable to ward off disruptions in their long-term relationship with their physicians, chances are increased that they will have to deal with an unresponsive system without the benefits of significant support from their physicians.

Even as we acknowledge here that having a long-term positive relationship with one's physician can help patients obtain more responsive care, we should also emphasize that even the most helpful and involved physician cannot guarantee responsive care at all times (Groopman, 2000). Breakdowns in responsiveness of care can occur on many levels in today's complex health care system. Thus, for example, even a vigilant, involved physician cannot always anticipate scheduling delays for lab work or for seeing specialists. Accordingly, patients are left to decide when assumptions of responsiveness by the health care system may be warranted.

B. Recognizing the Adaptive Tasks of the Patient.

The second component of proactivity in enhancing one's health care relates to identifying and assessing key adaptive tasks for those seeking to obtain health care. Identification of the adaptive tasks the patient must deal with represents the first step toward developing systematic approaches to increasing responsiveness of health care in each of these areas. Key adaptive tasks for patients are listed below.

Adaptive Tasks of Patient (and/or Health Care Partner) During an Illness Episode

(1) Gaining access to health care services
   • Scheduling appointments with Physician
   • Scheduling appointments for diagnostic tests
   • Scheduling appointments for treatments
   • Obtaining referral to specialists
   • Fostering communication between members of the health care team.

(2) Gaining access to information needed for proactive involvement in health care
   • Finding information on clinical status of patient (test results, specialist reports)
   • Finding information on progress of diagnosis and treatment
   • Finding information on prognosis
   • Finding information on diagnostic and treatment options
   • Finding information on medical advances presenting new options
   • Finding information about strengths and weaknesses of current treatment setting and about alternative treatment settings.

(3) Obtaining financing for health care
   • Dealing with insurance companies to approve health care expenditures
   • Dealing with bills and payments.

(4) Enhancing comfort and quality of life
   • Controlling pain
   • Controlling anxiety about illness and outcomes
   • Managing work and family obligations
   • Maintaining meaningful activities and relationships.

Being aware of the adaptive tasks faced in each situation encountered during an illness episode enhances patient competence and opportunities for proactivity (Kahana & Kahana, 1996). Based on our experiences and our observations during the three illness episodes we encountered, it was typical to confront multiple adaptive tasks simultaneously. A major issue faced by the patient and key family health care partners involved prioritizing which adaptive tasks should be handled first, and recognizing that patient priorities and the priorities of representatives of the health care system do not always coincide (Edgman-Levitan, 1996). During acute illness episodes the patient's major adaptive task may have involved dealing with discomfort and anxiety, while communication from physicians may have dealt with the need to obtain further diagnostic tests. In fact, key health care providers such as primary care physicians or nurses were not typically aware of the diverse adaptive tasks confronting the patient. At times the most important aspects of proactive adaptations were related to bringing to the attention of health care providers patient priorities about adaptive tasks to be dealt with. Rather than relying on responsiveness of health care providers, proactive consumerism calls for facilitating work of health care providers toward assisting the patient in meeting adaptive tasks.

In the area of gaining access to health care services, we found it helpful to inform physicians of our progress in scheduling appointments for diagnostic tests and took an active role in fostering communication between members of the health care team. Thus, for example we routinely tried to obtain copies of test results. Having test results in our possession allowed us to mail, fax or hand-deliver needed information to specialists. Additionally, having information on specific diagnostic data was also useful in accessing information about diagnostic and treatment options. We will provide more detailed illustrations of these principles as we discuss the proactive adaptations we found useful in our health care experiences.

C. Examining Assumptions about Responsiveness of the Health Care System

In order to optimize one's health care in today's complex system of care delivery, the consumer must start by examining and, where necessary, rejecting the use of traditional assumptions regarding responsiveness of the system. Our time-honored approach to getting medical care (diagnosis or treatment for serious illness) is to place ourselves in the hands of our doctor and the health care system and hope for the best. In doing so we make a whole series of reassuring but untested assumptions.

Common assumptions made by patients and families about medical care include:

- Recent medical advances related to our problem will be known by our doctor.
- Someone among our health care providers will research advances related to the unique or unusual problems posed by our case.
- Health care providers will inquire about patient preferences, goals, and values, and will consider these preferences in providing care.
- Patients will be given ready access to their medical records.
- Diverse doctors and health care providers will communicate with each other about our care.
- Our physician's recommendations for follow-up, diagnosis or treatment will be implemented in a timely manner.
- If the best state-of-the-art care is outside of the facility with which our physician is affiliated, we will be alerted by our physician about opportunities to select more experienced providers and facilities.
- There is reasonable uniformity across health care facilities in the level of training or expertise of ancillary staff (nurses, technicians, therapists) and equipment used in treatment and diagnosis is comparable across hospitals or outpatient facilities.
- We will be alerted by our physician to alternative approaches to the treatment of our problem, including those available in neighboring or even distant facilities.
- Waiting for a diagnosis or treatment poses few risks. Delays are based on lack of urgency of our needs rather than provider convenience.
- Important information which will help us cope with our illness will be provided to us about activity or lifestyle limitations, diet and nutritional needs, and social and psychological repercussions of our illness.

Our observations and experiences during our three illness episodes provided instances where each of these assumptions proved to be partially or wholly untrue. Our first illustrative example deals with our experiences regarding un-responsiveness of the system in providing access to the patient's health records.

Illustrative Example One – Facing Unresponsiveness in the Health Care System: Lack of Access to Health Care Records

The patient was hospitalized for surgery. His physician brother arrived for a brief visit and wanted to be in a position to assist and reassure the patient about his general medical status. The patient asked that his brother be permitted to see his chart. The nurses said this could be accomplished only with the treating physician's approval. The resident physician was paged and came to the patient's room, where he reassured both brothers that it was acceptable for the chart to be reviewed. After the resident left, the nurses would not produce the chart. They claimed the resident did not leave an order in writing. The resident could not be reached a second time and a nursing supervisor was called and refused to release the chart. She now claimed that the resident had no authority to release the chart to the patient for review, and that an application needed to be filed with the patient records office. Permission from this office takes several weeks to obtain, the patient was told. The patient tried to explain that his brother must return to a distant city and would not be in a position to review the chart weeks later. The nursing supervisor was annoyed and unsympathetic. The patient was discharged the same evening, and previously friendly and cooperative nurses in charge of his case scurried around speechlessly and provided him only with the most cursory discharge instructions. When his wife came to pick the patient up, she immediately sensed that something had gone wrong and the patient left with a sense of great fear and a feeling of having been viewed as a troublemaker.

Patient efforts to get the records released were ineffective in this instance because representatives of the system became more entrenched in their unresponsiveness when challenged. Challenging authority, based on our observations, is likely to increase unresponsiveness.

D. Using Proactive Approaches to Deal with Unresponsiveness in the Health Care System.

We now turn to discussing proactive adaptations which patients can use as they encounter unresponsiveness within the system. When we, as patients and patient advocates, recognize unresponsiveness in the system, we are faced with the questions: Are we helpless in our own care? Are we doomed to suffer adverse outcomes? Do we need to find a new doctor or health care provider?

In this section we suggest some approaches which can allow patients to find satisfactory answers to these questions through rejecting the specter of helplessness by taking responsibility, and by planning and by getting involved in their own care. Having recognized that their assumptions of responsive health care are not being met in given situations, patients must identify and address unresponsiveness in the system around each of their key adaptive tasks.

When we think of ways in which patients can cope with their illness in the context of the health care system, medical sociologists typically assume compliance to be the appropriate currency (Conrad, 1987; Eraker, Kirscht & Becker, 1984). However, the value of compliance assumes a generally responsive health care system. It clearly makes sense to expect that patients follow doctors' directives for taking medications, avoiding health risks, or coming for treatments. Such compliance also makes sense in the context of a caring, long-term doctor-patient relationship. We can safely assume that the physician in charge of our care makes recommendations knowledgeably and with the best interests of his/her patients in mind.

In the case of an unresponsive system, one can no longer assume that recommendations to the patient are being made with his or her best interests in mind. Typically, such unresponsiveness is not based on willfully ignoring the needs of the patient, but in the complex web of the system, people may have no clear knowledge of the patient or his or her best interests. It has been argued that multiple staff involved in clinical care, coupled with diffuseness of authority leaves patients with a sense that no one is fully in charge of their care (Gerteis et al., 1996).

In fact, the majority of the patient's interactions or encounters take place not with the doctor but with ancillary personnel such as appointment schedulers or lab personnel. In the era of managed care it appears, from the patient's perspective, that physicians have been forced to relinquish coordination of the patient's care to diverse support personnel (Gerteis et al., 1996). Accordingly, when a patient presents with an acute problem requiring diagnosis, the patient is typically sent to make arrangements for diagnostic tests with an anonymous clerk scheduling the appointment. When that scheduler suggests that the appointment take place a month or two after the call, compliance with this recommendation is seldom in the patient's best interests.

Waitzkin (1985) found that lack of accurate and complete information received from their doctors is the major source of dissatisfaction for patients. Giving patients access to their medical records is one of the suggestions for improving medical care in the framework of patient centered medicine (Stevens, Stagg & Mackay, 1977; Allshouse, 1996). Yet, in practice, patients face many barriers to gaining access to their own medical records. Patient access to information is limited by constraint policies, such as keeping patient records at the nursing station or restricting use of hospital libraries to medical staff (Allshouse, 1996).

So what can the patient do to deal with an unresponsive system? S/he must proactively consider changing the system to be more responsive, outwitting the system when efforts to make it more responsive fail, or leave the system when other solutions fail. We must honestly acknowledge that the patient and his or her informal health care partner cannot insure the patient's well-being without being willing to break out of the unresponsive mold and recognize that they must be the major advocates and agents to insure positive outcomes when the system is unresponsive. It is notable that several recent books by physicians who have personally encountered unresponsiveness in the health care system advocate assuming vigilance and proactivity as a means of enhancing outcomes of health care (Blau & Shimberg, 2000; Groopman, 2000). Some solutions may be achieved well within the bonds of the existing system, while others necessitate going around the system or breaking the system down into more manageable components. Our own health care experiences during the episodes we are reporting here yielded three useful proactive strategies in dealing correctively with unresponsiveness in the health care system. We will briefly review these strategies, but also acknowledge that they do not always succeed in overcoming unresponsive care.

II. Corrective Adaptations

A. Linking Up with Responsive People Behind the Unresponsive System

In considering proactive options for patients in dealing with unresponsiveness of the system, we must recognize that changing the entire system and its modus operandi do not constitute a realistic short-term goal. In any health care encounter the patient's goal is to elicit responsiveness to his or her immediate or ongoing needs. Unresponsiveness is often predicated on not seeing the patient as an individual with needs. The patient is simply lost in the shuffle. The patient must somehow make himself/herself become visible and noticed as an individual. Generally this goal is best accomplished by creating linkages or even bonds with the staff of the large impersonal and unresponsive system. The challenge to the patient and health care partner becomes to reach and solicit help from individuals who are the representatives of the "uncaring" system.

Our second illustration exemplifies a successful attempt to gain access to a needed diagnostic test in a timely fashion, by going around the bureaucracy and by linking up with individual workers within the system. This illustration also describes a health care challenge wherein physicians were totally unaware of the problem of delayed scheduling of diagnostic tests.

Illustrative Example Two. Gaining Timely Access to Diagnostic Tests

This incident relates to an acute illness episode involving the first author, which resolved itself satisfactorily. The patient developed acute visual problems which could not be diagnosed by ophthalmologists. She was referred by her internist for an MRI to diagnose the problem. She was handed a referral form and advised to contact the MRI facility for an appointment. The earliest appointment offered to her was more than one month away. She could not accept the assumption that her problem could wait that long to be diagnosed. She first considered recontacting the referring physician to see if he originally meant to order an urgent appointment or would at least request an urgent appointment now. However, the physician could not be readily contacted. Clearly the adaptive task at hand was to obtain an earlier appointment. A second call to the appointment desk did not yield much more success. A bureaucratic sounding voice at the other end of the line simply reiterated, "We are booked up for the next month." The person hung up before there was any time to plead the emergency nature of the problem. Evidently, solutions for working within the unresponsive system would not work. The patient, propelled by both fear and independence, decided to actually visit the MRI department where the tests were to be conducted. She observed the setting for a while and noted a friendly receptionist among those processing patients with prearranged appointments. She approached the receptionist and asked for her assistance. Describing her visual problem and concerns about her diagnosis, she asked if there were any cancellations so she could be seen without delay. Based on this face-to-face contact, the receptionist appeared to be sympathetic and volunteered to check if there was something she could do to assist. She returned shortly and reported that, while there were no cancellations for the same day, if the patient was willing to come in at 7 a.m. the next morning, they could fit her in. This was a most welcome resolution to the problem, and the patient expressed her gratitude and appreciation to the receptionist. Proactive efforts of linking up with individuals representing the health care system proved to help engender responsiveness in the system and help the patient gain access to the needed diagnostic services.

B. Gathering and Using Information About Options Proactively

Patients facing illness situations need a carefully executed information-gathering strategy. Major information sources include medical libraries, bookstores, and the world wide web. Information must be regularly updated as symptoms or conditions change, and as the patient obtains more information about likely diagnosis and treatment. The patient needs to find someone among his/her family, friends, or professional resources, who can access state-of-the-art medical information and provide feedback about the patient's understanding of the technical information retrieved. Patients need to find resource individuals who can translate medical terminology into simple English. Some medical centers (e.g. Massachusetts General Hospital) provide access to librarians who can assist in information retrieval.

The patient must also be well informed about the qualifications of his/her physician and other health care professionals. S/he needs to learn about degree of familiarity their doctor may have with their specific illness and related treatment options. In order to obtain needed information, the patient must be willing to take the risk of disclosing information about his/her health problems to those who can be of assistance. Sometimes this involves setting aside normal reticence or embarrassment to disclose personal information to others. Part of the proactivity needed in information gathering efforts involves recognition that health care providers or health care institutions are not always forthcoming in providing information about health care options available to the patient. At the same time, patients must also be cautious in evaluating how authoritative and current information sources might be. Nevertheless, the increasing availability of the internet puts the patient within easy reach of sophisticated information resources such as Medline databases and the resources of the National Library of Medicine.

To the extent that patients have a close and trusting relationship with their physicians, proactive information gathering may best be utilized by the patient by discussing with their physician the information gathered by themselves or their informal caregivers. Armed with needed information about options for meeting illness adaptive tasks, the patient and his or her informal health care partner must communicate successfully with representatives of the health care system to achieve desired health care goals. We provide an illustrative example of our successful use of proactive information gathering in finding an expert to perform surgery.

Illustrative Example Three. Successful Use of Proactive Principles in Finding an Expert to Perform Surgery

There was a diagnosis of a rare tumor found unexpectedly in a fit patient, previously in excellent health. The diagnosis was made by a specialist in a leading hospital, neighboring another world-class health center. Surgery was clearly indicated and described as both risky and extensive. The task at hand was finding the most experienced and competent surgeon to consult. Physicians at both local health centers were consulted and each recommended experts at their own or, in one case, at the neighboring hospital. The patient was cautioned about the extremely large expenses he may incur should the surgery be performed at a hospital other than the one covered by his insurance policy. Because the required surgery is infrequently performed, we attempted to query recommended surgeons about the frequency with which they have performed it. Responses were very general and ranged from someone at one of the recommended surgeon's offices assuring us that he does two of these general surgical procedures per week to assurances from another surgeon's office that he does many similar operations all the time. Our lingering unease about the lack of specificity and candor with which information was being given made us sense unresponsiveness and propelled us toward greater proactivity in our search for information. We took a two-pronged approach. With the assistance of a graduate student, we undertook a systematic library search on the required operation. We also enlisted the help of our adult children to contact acquaintances in the medical field who might make personal inquiries; thereby not limiting us to the local hospitals we had previously considered. The results of both information gathering strategies were unexpected but convergent.

Through an e-mail inquiry of our son's acquaintance, who is a surgeon in a different field, we were given two names of surgeons who were recognized as national experts in the needed surgical procedure. Our literature searches provided direct information from respected medical publications about the reduced mortality and morbidity experienced by patients undergoing this surgery at the most experienced centers which specialize in this procedure. The two centers we found in the literature were the same ones suggested by the acquaintance. Interestingly, we also found publications describing surgical experience with the procedure in question, from the local medical center. These publications revealed that only a very limited number of identical surgeries had been performed at that center.

We now needed to take the next step and consult one of the two experts we identified. Here we encountered overt discouragement from local physicians about obtaining care out-of-town, while at the same time we also experienced challenges in getting an appointment to see the out-of-town expert. A major deterrent to seeking help out-of-town was raised by our local physician, who, meaning well, cautioned us that aftercare may become problematic for those who have surgery at another location. We did nevertheless consult the out-of-town expert after our concerned, persistent, but friendly communications succeeded in yielding an appointment. We found our encounter with the surgeon specializing in the needed procedure to be a wonderful example of patient responsive care. Because of his team's vast experience, the surgery involved no stay in intensive care and was approached in a far more optimistic manner.

Further, we were reassured that aftercare in our home town should pose no obstacles, particularly since this renowned out-of-town surgeon had fine collegial relations with surgeons in each of the major medical centers in our home town.

What we had anticipated to be a difficult decision, turned out to have been made for us. Out of town patients were welcomed, and their stay was facilitated at this medical center, and the level of interest in our case far surpassed anything we had previously encountered. Indeed, the major surgery was successful and uneventful. We were particularly pleased that during the required hospital stay, members of the family could be with the patient around-the-clock throughout the period of post-surgical recovery. We were also pleased with the high level of familiarity of nursing staff and ancillary personnel with this type of case. Proactive and friendly communication with the physician, who originally discouraged our consulting out-of-town experts, facilitated our re-entry to care at home. We regularly e-mailed our physicians back home about progress of the surgery. Our initiative in keeping everyone "in the loop" resulted in being welcomed back warmly. In fact, one of the many physicians who had initially not supported our seeking help out-of-town noted, "I am glad you sought out the best person in this field. Had I had a similar diagnosis, I would also have chosen him to do my surgery."

This set of experiences reveals that our assumptions about the responsiveness of the health care system in helping the patient identify the best experts to do technically difficult surgery, were not supported. No one among the local specialists who were consulted raised concerns about limited experience by local medical centers in performing surgery, nor did anyone share information about shorter hospital stays and lack of need for a stay in the intensive care unit at more experienced centers. Adaptive tasks for the patient and health care partner included getting information, gaining access to nationally recognized experts and maintaining communication with local physicians. Proactive adaptations yielded potentially life-saving benefits and greatly reduced patient and family suffering.

C. Communicating effectively with Representatives of the System to Achieve Health Care Goals

Patients and their informal health care partners cannot assume that representatives of the health care system will freely and responsively communicate with either the patient, their family, or with one another. In fact, lack of good communication has been widely recognized to exist in modern health care delivery (Daley, 1996). Patients need to be clear about the nature of their illness, both of predictable and unpredictable aspects of their condition, as well as potential benefits and risks of treatments to be undertaken (Groopman, 2000; Kreps, 1996).

The patient must find creative ways to facilitate communication with and among health care providers. Strategies may range from inquiries and reminders to facilitation of communication among health care providers (e.g. patient volunteers to deliver x-rays to specialists by cab). Rather than being passive recipients of communication, proactive patients must initiate communication with nurses, physicians, and pharmacists. By writing down, prioritizing, and organizing one's questions, the patient and the informal health care partner can also minimize resistance to their proactivity and assertiveness. When the physician is curt, appears distracted, or disinterested, it is particularly important for the patient to find a diplomatic way to be heard. At times, we found it useful to make a second appointment in order to find the opportunity to discuss important issues. However, if the patient is not satisfied that (s)he was heard and the issues were fully discussed, second or even third opinions from other physicians should be sought, particularly when an illness is rare or life-threatening, and treatment options are unclear (Groopman, 2000).

An important objective of communication relates to the patient and his/her family members establishing good rapport with health care providers. Frequent communication to acknowledge and express appreciation for responsive care can serve to reinforce positive interactions, which, in turn, can enhance responsiveness of care for the patient during future illness episodes. Recognizing the key roles played by ancillary personnel, including receptionists, schedulers, lab technicians, and nursing assistants in patient care is important. The patient and family engaging positive and appreciative communications with representatives of the health care system can thus serve preventive as well as corrective functions (Kahana & Kahana, 1996). To the extent that patients and family members acknowledge providers as significant people and take note of their needs and perspectives, the likelihood of responsive care in the future is increased, both for themselves and for other patients.

Beyond the areas of divergence and clash in perspectives between patients and health care providers, notable areas of convergence also exist (Kahana et al., 1999). Hospital staff and patients share frustrations in being controlled by the hospital system, and experience areas of powerlessness relative to the system. Physicians and nursing staff may all find hospital food tasteless. Physicians and patients may share in the frustration of a long wait in transferring a patient from one unit to the next (Kahana et al., 1996b). To the extent that patients can identify these shared frustrations, and even discuss these with health care providers, an alliance can develop between patients and those who care for them, which can diminish frustrations and enhance responsiveness of care (Candib, 1987). It also must be remembered that both hospital staff and patients share a most fundamental similarity because they are all human. Ultimately, it is this shared humanity which provides the key to obtaining patient-responsive care.

Illustrative Example Four. Enhancing Responsiveness of Care Through Communication and Forging Alliances with Health Care Providers

Our next illustration describes success in enhancing responsiveness of care through communication and providing help to health care providers. During the recovery period after surgery, a home health care nurse assisted the patient. The patient and his health care partner established good communication with the nurse. Instead of focusing on the patient's needs, the patient and spouse acknowledged challenges faced by the nurse in taking care of her large and diverse patient load. They were empathetic to her as she ran late on two consecutive days and she reciprocated by sharing her own frustrations with increasing financial constraints and work loads within her home health agency. The next day, in visiting the patient, the nurse was self-disclosing enough to note that she was missing some supplies she needed to optimally care for the patient, and that she would have to make do with improvised equipment until a new set of supplies could be delivered the following week. Based on the partnership developed with the nurse, the patient's spouse felt free to volunteer to call the hospital where the surgery was done and speak to one of the charge nurses who had offered to assist if problems should arise. The home health care nurse was doubtful of the likely success of this initiative, but agreed to have the family proceed. The familiar charge nurse was on duty in the hospital when called and had the needed supply available for pickup within the next half hour. The patient's comfort level was much enhanced by the home care nurse using proper supplies. In this case, good rapport with health care staff, coupled with proactive initiatives to marshal support, and linking health care providers, served to enhance responsiveness of care.

CONCLUSIONS

This paper focused on the creative ways that consumers can enhance the quality of their health care through personal advocacy. Patient proactivity was discussed as a challenging but potentially very rewarding option in our increasingly complex and bewildering world of health care delivery. The proactive patient involvement in health care that we detail represents a paradigm shift for patients who have been generally expected to be passive recipients of care. We advocate taking responsibility at a time in patients' lives when illness strikes and when patients have traditionally expected to be taken care of. Systematic guidelines for proactivity may be particularly useful to patients and family members dealing with serious and chronic health care challenges.

In addition to the benefits of proactive consumerism insuring improved health care, such orientations also benefit the patient in other ways. Patients achieve a much needed sense of mastery and competence from enacting proactive adaptations at times when symptoms and prospects of serious illness may undermine their self worth. Demonstrating competence in approaching the health care system enables patients to counteract learned helplessness and contributes to learned optimism (Seligman, 1972). Positive affective states which result from proactivity may even have beneficial effects on resistance to illness and promote healing (Benson & Stark, 1996; Justice, 1998).

Proactive involvement in health care through building alliances represents an alternative avenue to consumerism than traditional expectations of confrontational or grievance-based coping strategies. Nevertheless, we retain the term "consumerism" in describing our approach because we believe that multiple approaches to proactivity in obtaining health care are possible. Alliances with the human beings who inhabit the bureaucracies which deliver unresponsive care must be creatively forged by patients and their advocates.

Proactive and competent coping by individuals ultimately can and should also be supplemented by changes brought about by health policy makers and health care practitioners which contribute to developing a more responsive system. Representatives of the health care system should also encourage proactive patient involvement in care, as it makes patients resourceful partners who can assist providers in their quest to deliver high quality health care.

REFERENCES

Abraham, L. K. (1993). *Mama Might be Better Off Dead: The Failure of Health Care in Urban America*. Chicago, IL: University of Chicago Press.

Allshouse, K. D. (1996). Treating patients as individuals. In: M. Gerteis, S. Edgman-Levitan, J. Daley & T. L. Delbanco (Eds), *Through the Patient's Eyes. Understanding and Promoting Patient-Centered Care* (pp. 19–44). San Francisco, CA: Jossey-Bass Publishers.

Benson, H., & Stark, M. (1996). *Timeless Healing: The Power and Biology of Belief*. New York: Scribner.

Blau, S., & Shimberg, E. (2000). *How to Get Out of the Hospital Alive: A Guide to Patient Power*. Edison, NJ: Castle Books.

Brennan, P. F. (1995). Patient satisfaction and normative decision theory. *Journal of the American Medical Informatics Association, 2*(4), 250–259.

Candib, L. M. (1987). What doctors tell about themselves to patients: Implications for intimacy and reciprocity in the relationship. *Family Medicine, 19*(1), 23–30.

Conrad, P. (1987). The experience of illness. *Social Science and Medicine, 24*(8), 700–701.

Daley, J. (1996). Overcoming the barrier of words. In: M. Gerteis, S. Edgman-Levitan, J. Daley & T. L. Delbanco (Eds), *Through the Patient's Eyes. Understanding and Promoting Patient-Centered Care* (pp. 72–95). San Francisco, CA: Jossey-Bass Publishers.

Edgman-Levitan, S. (1996). Providing effective emotional support. In: M. Gerteis, S. Edgman-Levitan, J. Daley & T. L. Delbanco (Eds), *Through the Patient's Eyes. Understanding and Promoting Patient-Centered Care* (pp. 154–177). San Francisco, CA: Jossey-Bass Publishers.

Eraker, S. A., Kirscht, J. P., & Becker, M. H. (1984). Understanding and improving patient compliance. *Annals of Internal Medicine, 100*(2) 258–268.

Flocke, S. A., Stange, K. C., & Zyzanski, S. J. (1997). The impact of insurance type and forced discontinuity on the delivery of primary care. *Journal of Family Practice, 45*(6) 129–135.

Gerteis, M. (1996). Coordinating care and integrating services. In: M. Gerteis, S. Edgman-Levitan, J. Daley, & T. L. Delbanco (Eds), *Through the Patient's Eyes. Understanding and Promoting Patient-Centered Care* (pp. 45–71). San Francisco, CA: Jossey-Bass Publishers.

Gerteis, M., Edgman-Levitan, S., Daley, J., Delbanco, T. L. (1996). *Through the Patient's Eyes: Understanding and Promoting Patient-Centered Care.* San Francisco, CA: Jossey-Bass.

Goldberg, C. (2000, June 12). Referendums from states are seeking to overhaul the health care system in the U.S. *The New York Times*, A1.

Groopman, J. (2000). *Second Opinions: Stories of Institution and Choice in the Changing World of Medicine.* New York, NY: Viking Press.

Hafferty, F., & Light, D. (1995). Professional Dynamics and the Changing Nature of Medical Work. *Journal of Health and Social Behavior, V*, 1995, 132–153.

Haug, M. R., & Lavin, B. (1978). Public challenge to physician authority: A comparative perspective and empirical test. Presented at the 1978 International Sociological Association.

Justice, B. (1998). *A Different Kind of Health: Finding Well-Being Despite Illness.* Houston, TX: Peak Press.

Kahana, E. (1982). A congruence model of person-environment interactions. In: M. P. Lawton, P. G. Windley & T. O. Byerts (Eds), *Aging and the Environment: Theoretical Approaches* (pp. 97–120). New York: Springer Publishing Company.

Kahana, E., & Kahana, B. (1996). Conceptual and empirical advances in understanding aging well through proactive adaptation. In: V. Bengtson (Ed.), *Adulthood and Aging: Research on Continuities and Discontinuities* (pp. 18–41). New York: Springer Publishing Co.

Kahana, E., Kahana, B., & Chirayath, H. (1999a). Innovations in institutional care from a patient-responsive perspective. In: D. Biegel & A. Blum (Eds), *Innovations in Practice and Service Delivery Across the Lifespan* (pp. 249–275). New York: Oxford University Press.

Kahana, E., Kahana, B., Kercher, K., & Chirayath, H. (1999b). Developing a patient-responsive model of hospital care. *Research in Sociology of Health Care, 16*, 31–54.

Kahana, E., Stange, K., Meehan, R., & Raff, L. (1997). Forced disruption in continuity of primary care: Managed care, the patient's perspective. *Sociological Focus, 30*, 172–182.

Kao, A. C. (1998). Trust and agency: the patient-physician relationship in the era of managed care. *Dissertation Abstracts International Section A: Humanities and Social Sciences, 59*(5A), 1790.

Kreps, G. L. (1996). Promoting a consumer orientation to health care and health promotion. *Journal of Health Psychology, 1*(1), 41–48.

Lerner, D., & Schwartz, C. E. (2000). Quality of life in health, illness, and medical care. In: C. E. Bird, P. Conrad & A. M. Fremont (Eds), *Handbook of Medical Sociology* (5th ed., pp. 298–308). Upper Saddle River, NJ: Prentice Hall.

Liang, B. A., & Cullen, D. J. (1999). The legal system and patient safety: Charting a divergent course: The relationship between malpractice litigation and human errors. *Anesthesiology, 91*(3), 609–611.

Lorber, J. (1975). Good patients and problem patients: Conformity and deviance in a general hospital. *Journal of Health and Social Behavior, 16*(2), 213–225.

Macklin, R. (1993). *Enemies of Patients*. New York, NY: Oxford University Press.

Mechanic, D. (1998). The functions and limitations of trust in the provision of medical care. *Journal of Health Politics, Policy, and Law, 23*(4), 661–686.

Moloney & Paul (1996). Rebuilding public trust and confidence. In: M. Gerteis (Ed.), *Through the Patient's Eyes. Understanding and Promoting Patient-Centered Care*. San Francisco, CA: Jossey-Bass Publishers. New York Times, 2000.

Quint, J. C. (1972). Institutionalized practices of information control. In: E. Freidson & J. Lorber (Eds), *Medical Men and Their Work*. Chicago, IL.: Aldine Publishers.

Pescosolido, B. A., Wright, E. R., McGrew, J., Mesch, D. J., Hohmann, A., Sullivan, W. P., Haugh, D., DeLiberty, R., & McDonel, E. C. (1997). The human and organizational markers of health system change: Framing studies of hospital downsizing and closure. In: J. J. Kronenfeld (Ed.), *Research in the Sociology of Health Care* (pp. 69–95). Greenwich: JAI Press Inc.

Rosenthal, M. M. (1999). Medical uncertainty, medical collegiality, and improving quality of care. *Research in the Sociology of Health Care, 16,* 3–30.

Seligman, M. (1972). Learned helplessness. *Annual Review of Medicine, 1972,* 207–412.

Stevens, D. P., Stagg, R., & Mackay, I. R. (1977). What happens when hospitalized patients see their own records? *Annals of Internal Medicine, 86,* 474–477.

Stewart, M., Brown, J. B., Weston, W. W., McWhinney, I. R., McWilliam, C. L., & Freeman, T. R. (1996). *Patient Centered Medicine. Transforming the Clinical Method*. Thousand Oaks, CA: Sage Publications.

Waitzkin, H. (1985). Information giving in medical care. *Journal of Health and Social Behavior, 26*(2), 81–101.

Wholey, D. R., Padman, R., Hamer, R., & Schwarz, S. (2000). The diffusion of information technology among health maintenance organizations. *Health Care Management Review, 25*(2), 24–33.

CONSUMER DESIRE AND MEDICAL PRACTICE

Barbara Hanson

ABSTRACT

This paper presents the construct of "medicalism" – a widespread mind set for defining and dealing with troubles – as a theory of how health service provision interacts with the desires of consumers. Medicalism has three components: individuation, externalization, and just pain. In concert this sets up a situation where explanation may be sought more fervently than cure. Thus, failure, rather than success, assures escalating demand for health services. This suggests thinking about how health care fits into the full array of defining and dealing with troubles. Medicine and war can be seen as two sides of the same coin, processes that individuate, and externalize troubles while justifying pain or expense. Escalating health care costs can then be seen as related to the refusal or inability to deal with problems in other spheres. Health care becomes a dumping ground for a vast range of social problems ranging from poverty to sexual assault. In so doing, health care is forced to do less of what it does well and more of what it does badly.

CONSUMER DESIRE AND MEDICAL PRACTICE

Medical intervention prospers because consumers wish it worked and get initial relief through the externalization of troubles. Therefore, consumer desire for

Changing Consumers and Changing Technology in Health Care and Health Care Delivery,
Volume 19, pages 45–57.
2001 by Elsevier Science Ltd.
ISBN: 0-7623-0808-7

success and relief, needs to be part of models of health care in order to recognize its interactive, co-emergent nature. Medicalism is a new model which does this by theorizing a world view or epistemological stance for dealing with troubles which is shared by care providers and consumers. It has three parts – individuation, externalization, and just pain.

The concept of medicalism can be used to understand and provide a context for current debates on the nature and funding of health care in North America, especially the public/private pay debate. Medicalism provides a way of explaining why demand for medical services is infinite. Because demand is infinite there will always be a movement to allow those who can pay to get more and have more say in what they get. This demand will also drive health service provision farther away from the need to be effective. That people want to believe it works is enough to perpetuate demand. It may even be the case that demand is most magnified in situations where effectiveness is least evident. This is exacerbated by changing technology for consumers,especially greater access to medical research information and online support networks via the internet and attention to genetics.

Medicalism goes beyond critique of medicine and models of medicalization by looking at a widely based contextual mind set for dealing with troubles of various kinds (Hanson, 1997), continuing similar work (Zola, 1972). It captures how troubles in general are seen and acted upon. It also questions which troubles are dealt with by which kind of intervention: medical as opposed to military, education, welfare, or other. This accounts for why the demand for health care, as one possible way of dealing with troubles, continues to grow.

The process of individuation, externalization, and pain justification allows troubles of varying sorts to be classified as health issues when they may be rooted in systemic issues like inequality. Seeing hyper activity, depression or stress, as troubles that can be managed with pharmaceuticals directs attention away from increasing class sizes in the public education system owing to under funding. Targeting the "Alzheimer's gene" neglects seeing the high association between old age and poverty, and the issues raised by the need for caregiving and social support (Wuest et al., 2001). The permeation of medicalism into a continually wider range of troubles is suggested by a recent article on "Anti-bullying interventions at school" in a health promotion journal (Stevens et al., 2001). Thus, I argue for attention to what is relegated to health care to fix and where the metaphors, and paradigms of medicine are used outside health care in realms such as social, defense, and economic policy.

This new model draws on and advances contemporary theories of medicalization (Conrad & Schneider, 1992) by moving to a model of pattern rather than cause. Greater efficacy is attributed to health care consumers by recognizing that they actively seek, finance, manage, and maintain, clinical contacts. Unlike previous models, it accounts for consumer spending on uninsured health products and services to the extent of a multi-billion dollar industry. Likewise it explains why consumers seek intervention, even repeatedly, in the face of failure, and with major iatrogenic effects.

COMPONENTS OF MEDICALISM

Medicalism as a mind set for dealing with troubles has three parts – individuation, externality, and just pain. Intertwined in actual events, they can be separated conceptually to shed light on how the process of dealing with troubles plays out. In total they present a model that can explain the perpetuation of medicine as a means of dealing with troubles even, perhaps especially, in the face of its failures.

Individuation

Individuation refers to the tendency to break down troubles into their components in order to figure out what is wrong, like taking apart a radio. In the frame of medicalism in health this most often means focusing on the body as the unit. Skin becomes the boundary within which problems occur, are caused, and thus where they can be treated. This division of troubles into bodies is a first step in finding the biological components of a particular problem. In health, individuation is a pre-requisite to biological reductionism, focus on the biology of bodies.

Reduction to parts involves trying to find a body, or person, to blame for a particular trouble. This was seen vividly in the build up to the Persian Gulf War in 1991 which repeatedly portrayed Saddam Hussein as "the problem" often drawing the parallel to Hitler. It has also surfaced in ill-fated attempts to try to explain distress in intimacy like anorexia, schizophrenia, incest, or violence, by pointing the finger at one person or group of people. The failure of this strategy led to the development of family therapy with its landmark journal *Family Process* which focused on contexts rather than individuals.

Individuation as a component of medicalism means that when a trouble is examined the tendency will be to isolate cause within an individual body or part of a system. This is a necessary first step on the way to blaming an individual component for a trouble followed by fixing or eliminating the errant component to make the problem go away. Individuation can be thought of broadly in terms of not just persons but also groups, institutions, policies, corporations, or agencies. In this sense the process of individuation can be seen as a search for a functional body, component, in which to isolate cause, attribute blame, then fix. It is an attempt to isolate cause and freeze process, in order to understand it. In medicine this often means sampling a piece of the alleged problem, taking a PAP smear, blood, urine, etc. Then the sample is put under a still viewing lens. Doing this does however, presume that troubles are detectable in components and the stopped single picture represents the process.

Medicine itself has recognized this problem for some time and moved to diagnostic tests like barium enemas and angiogram that look at the body part in process rather than the still time frame slice of an x-ray. Social science seems to have been slower to respond in that single slice time diagnostics are often relied on. Even where multiple time slices are used, the process must be inferred, it is not actually seen. Statistics like regressions, time series, phrase and gesture coding, must first take apart interaction, measure each answer, every six months, or each part of a sentence. This means that the whole process must be inferred by summing its parts.

To individuate is to define boundaries at the edge of a component. In health this often means using skin as a boundary. In other areas the components are different– a region of the world, a market, an oil company, an ideology etc. Common to all searches within the frame of medicalism is the search for a body, be it actual or functional, to define as the unit. This is a precursor to isolating troubles within this unit.

Externalization

Externalization is closely linked to individuation in that seating troubles in components is also a process of separating problems from context. Focus goes away from the system and onto one of its components. Troubles become externalized because they are outside the viewing lens.

Here the metaphor of disease is particularly relevant. Attributing causal status to an entity that is outside the realm of control shifts responsibility for troubles. This aspect of sickness is seen in Parson's conception of the sick role whereby the patient is not to blame for being sick, but is responsible for wanting and trying to get well (Parsons, 1979). It echoes in recent discussion of how illness is being constructed outside the body (Armstrong, 1995). This surfaces in attempts to try to come to grips with troubles like eating disorders, alcoholism, chronic pain. or depression, by attributing them to some medical illness – it's a disease. Regardless of the fact status of claims to some physical medical entity, the process of seeking such status is worthy of attention in its own right.

Why is it important to seek a disease label for these troubles? I argue that it is a need to absolve the individual for responsibility via externalization. By setting the trouble into a medical category it removes culpability, the person can't be blamed for it. It also opens up a means of action, seeking cure, that raises the status of the person in that they are working on, fixing something that is not their fault. Disease often absolves the person of guilt initially even though in the long run it opens the door to constructions of blaming the person through lifestyle.

The value of illness as a relief, and as such a reward, in seeking illness labels may come from the desire to externalize troubles. People may feel a tremendous relief when they find out that something is not their fault or there is nothing they can do in the case of incurable illness. This is echoed in the use of metaphors like "survivor", "victim" that connote attack, war, retaliation, wounding, and crime. There is absolution both from guilt for causing something and need to do something about it. Or, there may be a clear course of action even if it means giving up hope of cure.

Throughout my work on senile dementia (Hanson, 1991, 1989), I recall hearing how relieved families were when a diagnosis of senile dementia or Alzheimer's was attached This was the case with Rita Hayworth's daughter. After years of problematic behaviour, and diagnoses that attributed it to alcoholism, a new doctor gave the diagnosis of Alzheimer's. The biographer writes that Hayworth's daughter said there was a " 'sense of relief' in finally knowing the source of Rita's recent problems. 'So much embarrassment and heartache could have been saved', Yasmin declared, 'if at that time it had been known that Rita Hayworth was ill and not guilty of any misconduct" (Leaming, 1989, p. 330). Regardless of the actual physical relation of the illness to the behaviour, I argue that the process of defining as ill offers a certain benefit through absolution, even if only at first.

This is the consumer, seeker, side of the process. People seek medical labels because they offer the relief of externalization by absolving the person from wrongdoing. Others have observed this phenomena from the standpoint of social control in a pivotal text aptly titled "From Badness to Sickness" (Conrad & Schneider, 1992). I concur but add that this in not a one-way process. It continues because people may want to surrender control in order to excise the trouble, remove blame, and erase guilt.

Externalization likewise absolves the clinician in that they don't have to succeed in cure, they just have to try. Disease, if considered beyond their control, is not their fault. This may be a physical reality or limitation of current wisdom. Or, it is possible, that the process of externalization itself leads to inattention. The assumption that nothing can be done makes it functionally true in that people, both patients and clinicians, don't bother to look for alternative solutions. In this sense disease conceptions themselves might be the panacea of medicalism. They provide universal relief from the tension of search for explanation. In this sense it may be that the goal of medicalism is not cure but explanation.

This becomes interesting for the question of medicalism as a mind set for dealing with troubles within and outside medicine. Within medicine externalization can be seen as a crucial component of recent shifts in attention to lifestyle medicine – scolding people for smoking, being overweight, or not exercising. This expands into the emphasis on epidemiology that focuses on regional variations attributed to cultural differences in consumption or exercise. It is echoed in debates about whether smokers or obese persons should be given organ transplants. Other debates revolve around blaming patients for soaring health care costs because they use too much (Sweet, 1991). What this does is shift focus away from the efficacy of health care while maintaining the primacy of medical models for dealing with troubles in a medical frame.

This is where medicalism becomes self sealing. Its ineffectiveness for dealing with troubles internally leads to externalizing and shifting responsibility onto individuals and away from social context, while simultaneously maintaining control. It may even be the case that the less the likelihood of legitimate explanation the greater the attempts to seek explanation. I have explored this in depth elsewhere in terms of the search for genetic explanations (Hanson, 1995). This is a perfect set up for continuing the prosperity of medicine. It can never be blamed for what it can't fix but at the same time collects fees for blaming. In this sense its failings are as profitable, and perhaps even more, than its successes. The relevance of this view is suggested by a recent article offering that public spending on health seems to have little effect on health (Filmer & Pritchett, 1999).

Just Pain

Just pain refers to the notion that pain is right, proper, fair, well-founded, or fitting. It is OK for pain to occur in dealing with troubles. At times this is expanded to mean that pain itself indicates progress, success, or righteousness. "No, pain, no gain". I sense this popular appeal is at the heart of a recent advertisement for cough syrup: "It tastes awful and it works". I am using the term pain broadly to encompass not just the individual experience of pain but the pain of damage and spending money as well.

Just pain is tied to the allopathic notion that health and illness are opposites. Health is good, illness bad. Illness is killed or maimed to promote health. The goodness of health justifies the destruction of illness often regardless of cost, pain. This does however, presume that illness is bad and this badness justifies the cost of killing it. Medicalism can therefore, offer an explanation for why it was that allopathic medicine, among the range of possible alternatives, became prominent in North America (Clarke, 1990).

Just pain is reflected in the notion of collateral damage, that which must be destroyed in the course of eliminating some illness. In medicine this refers to what may be justly destroyed or disrupted in the course of treating some problem such as loss of 20% of one's liver, or a breast, to get out a malignant growth. This is crucial to the concept of medicalism in that surgical bombing and collateral damage surfaced as important metaphors in the Persian Gulf War. Surgical bombing became the term for killing or excising the illness. Collateral damage came to denote the many thousands of civilian casualties.

What is important for the concept of medicalism is seeing how inextricably intertwined the metaphors of medicine are in a broad range of ways of dealing with troubles. Justifiable pain, cost, suffering, is defined in terms of the intention to do good against evil. This is crucial in that just pain is defined by the intent to do good against evil with no necessary assessment of whether it is cost efficient. Is the degree of pain justified by the level of cure?

This comes through in health in that relative or absolute weighting of cost rarely comes into play. The massive cost of treatments like organ transplants, or vascular (usually cardiac) surgery, is rarely compared to is relative success. Or, as one journalist pointed out many medical procedures such as angiogram, heart bypass, or transplants, would never be approved if they were drugs, in that their relatively low success rates would not justify their level of risk (Whitaker, 1992). This lack of relative assessment transfers into the realm of defense in that the cost of war is seen as self justifying.

Just pain means that intervention is worthy in itself even if not successful and very costly. Doing something is better than doing nothing. This is not always the case, but is the justification sometimes. Further, relative valuing of intervention is becoming more and more prevalent in more patient centred clinical settings and particularly evident in feminist institutions like women's health centres and Toronto's world-renowned Women's College Hospital. Appropriate to my argument is the observation that this particular hospital is now fighting for its life in the context of neo-conservation anti-woman political agendas that dominate the current government. Since writing the first draft of this paper, Women's College Hospital has been forced to merge with a more traditional acute care institution. I argue that it may not be just the care philosophy itself that is so much under attack but the repeated evidence that patient-centered care works better and is more cost efficient.

This is not to be seen as blanket statement that medical action always or usually follows the course of just pain. It does, however, represent the tendency for this type of thinking to lead to cost and pain. Medicalism spawns a process that begins with individuation and seeks externalization of troubles. Because of this, understanding or explanation may be sought more fervently than cure. Creating an illness itself serves the process and provides rewards for all involved. Therefore, the process is served by categorization of troubles into an illness label. It need never cure. Defining and labelling is sufficient to ensure that the process with continue.

MEDICALISM IN THE CARE PROCESS

Medicalism as an epistemology reflects a desire to remove troubles from immediate responsibility by labeling them as pathological. This is relieving at first because it allows separation of "me" from "it", the person from the disease. The person can thus become part of the forces of good fighting evil. However this initial relief is fleeting since the individual is effectively removed from his or her body – a physical impossibility as far as I know. Goffman pointed out the dilemmas this creates in his discussion of the "Tinkering Trade" whereby the person must accompany the body in order to have it fixed (Goffman, 1961). The paradoxes of a "me"/"it" separation haunt a person in the illness process who is living in a body which they no longer own or control.

Lifestyle, and prevention metaphors with their accompanying blame become inevitable at this point. This comes out in neo-conservative political agendas which emphasize prevention in order to justify cutting social and health services (Roberts, Smith & Bryce, 1993). It is reflected in primary health care discourse (Johanson, Larson, Salo & Svardsudd, 1995) and a shift in epidemiological writing from cause to "risk" (Skolbekken, 1995). However, one source notes the tendency for doctors to espouse prevention but not take it seriously in actual practice by relegating it to a low status member of the team (Williams & Calnan, 1994). Prevention and lifestyle talk is good for the health care business, and for political agendas that aim to cut public services. It is crucial shift from people *having* something wrong, to *doing* something wrong.

Patients initiate, manage, and maintain the intervention process, seeking its positive aspects. Because of this, they need to be modeled as active agents in an interactive process (Thoits, 1994). I became sensitive to this possibility, patients seeking medical definitions as a more comfortable social category for themselves, during my work on nursing homes (Hanson, 1985). The people occupying beds in the nursing home I studied were supposed to be called "residents" rather than "patients", as part of a move to "de-medicalize" these institutions. Having nurses wear regular street clothes rather than uniforms was also part of this agenda. Staff took it as a sign of positive adjustment when a person being cared for would call it "home" rather than "hospital". However,

I observed that the people being cared for in the nursing home preferred medical definitions, wanted to be called patients, and started a successful movement to get nurses wearing uniforms and graduation caps. Further, they carried pills to recreation events where they were displayed on the table in front of them, such pills being unimportant by definition since all important medication was held by the nurses. These observations suggested that clients wanted medical definitions and actively sought them. To assume that nursing home clients are passive, have medical definitions forced upon them, or that medical labels are wholly negative seemed inappropriate. Instead I came to see health and illness as a concert with all parts contributing to the whole.

Support for my argument is found in Ford's example in women's health where women who in spite of knowing about and often living through the disasters of thalidomide, DES, and the Dalkon Shield, "want to believe that drugs and devices for women are safer today than they were 25 years ago" (Ford, 1986, p.28). Patients may want to be patients, and to believe that intervention will work. Medical seekers are, therefore, of equal causal status as medical providers. Medicalism presents a model whereby health care seeking and heath care provision are of equal causal status, co-emergent.

Medicine will continue because there is relief provided in defining things as illnesses, whether they are cured or not. The process is sustained not by cure but by the containment of fallibility. Attaching illness definitions may be more important than fixing problems because it removes the problem from the responsibility of both consumer and care provider. Evidence for this process is found in Ehrenreich and English's analysis of women's place in the healing business, that chronicles how lower class use of midwives became defined as a problem in order to justify fee collecting by one particular group of healers (1973). In Victorian times "heroic", often a metaphor for dangerous and painful, measures were done to justify fees to the consumer (Mitchinson, 1991). When illness becomes a thing or object it can be approached from a position of separation, indefinitely for the care provider, and at least initially for the consumer.

Medicalism differs from medicalization on several fronts. Medicalism is a mind set or epistemology while medicalization is a process or product which may grow out of this way of seeing. Medicalism relies on a cybernetic (feedback) rather than linear notion of causality. Medicine and medicalization are an offshoot or surface projection of the mind set of medicalism. In a model of medicalism, medicine and medicalization have no causal status separate from consumers' hope that it will work and moves to seek it.

The negative effects of medicalization (Conrad & Schneider, 1992), such as illness careers, or female pathologizing (McKinlay, 1972; Vertinsky, 1990; Hicks, 1994) may be more usefully modeled by seeing, through a model of medicalism that seeking and maintaining illness labels is a good/bad paradox. Like many narcotics, the highs that draw people to use them have consequent lows. Patients however do seek such highs. Further, patients seek clinicians rather than the reverse. A systems view means seeing that patients are not passive, they seek cure or at east explanation. This goes beyond the "medicalisation-demedicalisation" debate which permeates current thinking about health (Lowenberg & Davis, 1994).

DISCUSSION: MEDICALISM AS A CASE FOR MINIMALISM

Individuation, externalization, and pain justification, as intertwined components of the mind set of medicalism, are self sealing, hence self perpetuating. Even though medicine may not work or make things worse, the desire for it to work, and externalize troubles sustains it. This means that health intervention is prone to doing something rather than nothing and doing more rather than less. Specific health goals may, therefore, be better served by doing less rather than more, and nothing rather something. As the example of Canada's Women's College Hospital suggests, greater effectiveness and less costs may be looked at as undesirable. This principle could also have been at the heart of the defeat of health care reform in the U.S. in that reeling in the largest health care industry in the world would violate the notion of just pain that ensures an infinitely growing market for health care services.

One aspect of medicalism is the unlikelihood for a null or non-illness state. Diagnosis generally follows a pattern of ruling out alternatives. The process begins with either a report of a problem from a patient or detection of a problem by a clinician. The problem requires explanation. Information gathering, and testing will continue until an explanation is found. To wit, the case of Senile Dementia/Alzheimer's Disease, where diagnosis involves ruling out alternative explanations for the problem. When no evidence of anything else is found, the diagnosis of Senile Dementia/Alzheimer's is attached (Hanson, 1994). In this instance the attribution of a diagnosis requires only two things: (1). a report of a problem, and (2). a lack of evidence. Thus, even ostensibly healthy patients, ones with no observed abnormalities, can end up labeled as sick. Health may be a more elusive entity that illness.

Medicalism suggests looking at which troubles end up being relegated to health care in the full range of public policy alternatives. Emphasis on health care interventions could be indicative of a mode for not dealing with troubles by making them insolvable and self magnifying. Health care is often seen as a good thing, therefore a safe place to push for reform.

Health improves as socio-economic status improves. This is one of the most enduring and widely applicable findings in health research (Lantz et al., 2001). Health care, expensive though it is, may be much less expensive and contentious than full scale equity programs like universal socialized day care, equal access to high quality education, employment equity, or global wealth redistribution.

The widespread use and popularity of alternatives to bio-medical practices like supplements, exercise, spiritual growth, massage, chiropractics, has created an even greater total demand. To the allopathic principle of justified pain, alternatives offer that even if it doesn't work, it doesn't hurt. So, although there is no physical pain or detriment to health, there is still pain defined broadly in the sense of expense. It can even be argued that alternatives contribute more to the notion of trouble definition since they target lifestyle practices. We have prescriptions because we are sick and supplements to make or keep us well. Many alternative practices allow consumers to diagnose and treat themselves. This makes for more medicalism, even in the face of less bio-medicine.

Medicalism as a theoretical model goes deeper into health and illness as a question of individuation, externalization, and justifying pain (temporal, emotional, or financial). Whether a practice is bio-medical or alternative (wellness) does not change the fact that these are key elements of dealing with troubles that direct actions in a particular way, toward more. The alliance of bio-medical and alternatives is suggested by their increasing coalition: increasing insurance coverage for alternatives, teaching of alternatives in medical schools, practitioners holding credentials on both sides, and physicians recommending increased use of dietary supplements.

This makes it possible to consider that less might be better. I argue that we need to focus on effectiveness rather than amount and whether a trouble is best addressed by health care or another kind of intervention. This is particularly important as products, bio-medical, pharmaceutical, supplemental, and herbal, are being provided for a variety of troubles like stress, weight control, hangovers, depression, hyper-activity, or decreased memory. By individuating these troubles attention is directed away from systemic issues. Troubles may relate to bad relationships, discrimination, destructive work environments, unemployment, under-employment, poverty, debt (Drentea & Lavrakas 2000), chronic under funding of public education, or systemic inequalities for women (Akukwe, 2000). Focusing on things that people swallow as a way to address these troubles externalizes responsibility.

As I write this I realize that my arguments could be used to justify curtailing health care expenditure, cutting back on service provision. I prefer to think it of as a way of looking at medical services in terms of their actual benefits as weighed against the possibility that relegating troubles to health care may be a way of not dealing with them. In so doing, blame and responsibility are focused away from larger social issues like inequality.

It is time to carefully consider what health care best deals with rather than let this type of response to troubles grow indefinitely as more and more troubles end up in its purvey. I believe this is at the heart of the drive for genetic explanations and the willingness to direct massive funding to the human genome project. If the purvey of health care grows indefinitely so will its costs. This will limit its ability to deal with issues that are most appropriately dealt with as health issues.

ACKNOWLEDGMENTS

Many thanks to Georgina Feldberg for helpful comments on an earlier version of this manuscript. This paper emerges from a project which received support through Specific Grants, York University, Grants in Aid of Research, Atkinson College, and my appointments as Visiting Scholar, Institute for Research on Women, Rutgers University and Visiting Fellow, Department of Sociology, Princeton University.

E-mail: hansonbg@yorku.ca, Mail: Sociology-Atkinson, York University, 4700 Keele Street, Toronto, Ontario, Canada, M3J 1P3.

REFERENCES

Akukwe, C. (2000). Maternal and Child Health Services in the Twenty-First Century: Critical Issues, Challenges, and Opportunities. *Health Care for Women International, 23,* 641–653.

Armstrong, D. (1995). The Rise of Surveillance Medicine. *Sociology of Health and Illness, 17,* 393–404.

Clarke, J. N. (1990). *Health, Illness, and Medicine in Canada.* Toronto: McClelland & Stewart Inc.

Conrad, P., & Schneider, J. W. (1992). *Deviance and Medicalization: From Badness to Sickness.* Philadelphia: Temple University Press.

Drentea, P., & Lavrakas, P. J. (2000). Over the limit: the association among health, race and debt. *Social Science and Medicine, 50,* 517–529.

Ehrenreich, B., & English, D. (1973). *Witches, Midwives, and Nurses: A History of Women Healers.* New York: The Feminist Press.

Filmer, D., & Pritchett, L. (1999). The Impact of Public Spending on Health: Does Money Matter? *Social Science & Medicine, 49, 1309–1323.*

Ford, A. R. (1986). Hormones: Getting Out of Hand. In: K. McDonnell (Ed.), *Adverse Effects: Women and the Pharmaceutical Industry* (pp. 27–49). Toronto: Women's Educational Press.

Goffman, E. (1961). *Asylums.* New York: Doubleday and Company.

Hanson (1985). Negotiation of Self and Setting to Advantage: An Interactionist Consideration of Hursing Home Data. *Sociology of Health and Illness, 7,* 21–35.

Hanson (1989). Definitional Deficit: A Model of Senile Dementia in Context. *Family Process, 28,* 281–289.

Hanson (1991). Parts, Players and 'Patienting': The Social Construction of Senility. *Family Systems Medicine, 9*, 267–274.

Hanson (1994). 'Beyond Biologizing': The Unit Question in Gender Analysis of Senile Dementia. *International Journal of Sociology of the Family, 24*, 57–68.

Hanson (1995). See 'Genetic' Read Inexplicable: Genetic Explanations as a Means of Avoiding Responsibility while Justifying Profit and Control. *Canadian Sociology and Anthropology Association*, Montreal.

Hanson (1997). *Social Assumptions, Medical Categories*. Greenwich: JAI Press Inc.

Hicks, K. M. (1994). *Misdiagnosis: Woman as a Disease*. Allentown, Pennsylvania: People's Medical Sociey.

Johanson, M., Larson, U. S., Salo, R., & Svardsudd, K. (1995). Lifestyle in Primary Health Care Discourse. *Social Science and Medicine, 40*, 339–348.

Lantz, P. M., Lynch, J. W., House, J. S., Lepkowski, J. M., Mero, R. P., Musick, M. A., & Williams. (2001). Socioeconomic Disparities in Health Change in a Longitudinal Study of U.S. Adults: The Role of Health-Risk Behaviors. *Social Science & Medicine 53*, 29–40.

Leaming, B. (1989). *If this was Happiness: A Biography of Rita Hayworth*. New York: Balantine Books.

Lowenberg, J. S., & Davis, F. (1994). Beyond Medicalisation-Demedicalisation: The Case of Holistic Health. *Sociology of Health and Illness, 16*, 579–599.

McKinlay, J. B. (1972). The Sick Role – Illness and Pregnancy. *Social Science and Medicine, 6*, 561–572.

Mitchinson, W. (1991). *The Nature of Their Bodies: Women and Their Doctors in Victorian Canada*. Toronto: University of Toronto Press.

Parsons, T. (1979). *The Social System*. London: Routledge and Kegan Paul.

Roberts, H., Smith, S., & Bryce, C. (1993). Prevention is Better . . . *Sociology of Health and Illness, 15*, 447–463.

Skolbekken, J. (1995). The Risk Epidemic in Medical Journals. *Social Science and Medicine, 40*, 291–305.

Stevens, V., De Bourdeaudhuij, I., & Van Oost, P. (2001) Anti-Bullying Interventions at School: Aspects of Programme Adaptation and Critical Issues for Further Programme Development. *Health Promotion International, 16*(2), 155–167.

Sweet, L. (1991). Patients blamed as Medicare bills soar. *Toronto Star*, (1991), A11.

Thoits, P. A. (1994). Stressors and Problem-Solving: The Individual as Psychological Activist. *Journal of Health and Social Behavior, 35*, 143–159.

Vertinsky, P. (1990). *The Eternally Wounded Woman: Women, Doctors and Exercise in the Late Nineteenth Century*. New York: Manchester University Press.

Whitaker, J. (1992). Treatments that are More Lethal than the Diseases they Treat. *Wellness Today*, (December), 10–11.

Williams, S. J., & Calnan, M. (1994). Perspectives on Prevention: The Views of General Practitioners. *Sociology of Health and Illness, 16*, 372–393.

Wuest, J., King Ericson, P., Noerager Stern, P., & Irwin, G. W. (2001) Connected and Disconnected Support: The Impact on the Caregiving Process in Alzheimer's Disease. *Health Care for Women International, 22*, 115–130.

Zola, I. (1972). The Concept of Trouble and Sources of Medical Assistance – To Whom One Can Turn, with What and Why. *Social Science and Medicine, 6*, 673–679.

USING SOCIAL NETWORK PRINCIPLES TO STRUCTURE LINKAGES BETWEEN PROVIDERS AND PATIENTS

Neale R. Chumbler and James W. Grimm

ABSTRACT

This paper uses social network principles to explain the sources of and variations in relationships among health care providers and patients. General principles of social networks applied are global structure, structural equivalence, structural conduciveness, and the duality of network linkages. Specific principles employed to understand structural variability in provider interrelationships are structural encapsulation and structural excludability, centrality and integration, subgroups and structural holes, close ties versus weak ties, and the virtual ties created by computer-supported social networks (CSSNs). Various ways that social network principles help explain the evolving complexities of interconnectedness among health care providers and patients are demonstrated. Practical advantages of using social network principles to organize and to manage interrelationships among health care providers and patients are discussed.

Changing Consumers and Changing Technology in Health Care and Health Care Delivery,
Volume 19, pages 59–80.

INTRODUCTION

Dyadic patient referrals among physicians and between physicians and non-physician clinicians (e.g. podiatrists, physician assistants, nurse practitioners) have been the focus of several studies over the past few decades (Begun, 1978; Chumbler & Grimm, 1995; Kerssens & Groenewegen, 1990; Ritchey et al., 1989). However, more structurally complex, new approaches to conceptualizing and studying provider interrelationships are necessary now. A greatly increased number of patients including many who are chronically ill with multiple diseases are now treated by interrelated clusters of physicians as well as by non-physician clinicians (NPCs) (Cooper et al., 1998; Hafferty & Light, 1995; Sofaer, 1998). This paper intends to address the need for more complex ways to consider the emergence and structure of health care providers' interrelationships with patients in order to better address and deliver treatment for multiple health problems. The purposes of this paper are two-fold: (1) to show how the principles of social networks can be used to conceptualize interrelationships among larger clusters of health care providers, and (2) to show how a social networks framework can be used to better understand and construct health care arrangements among providers in order to address patients' needs more effectively and comprehensively.

General Principles

Social networks are defined as the enduring interrelationships among people or organizations that involve direct or indirect interpersonal connections (Burt, 1992; Carruthers & Babb, 2000; Cook & Whitmeyer, 1992; Granovetter, 1973, 1983, 1985; Marsden, 1983). Sociological research on social networks has focused on the nature of interpersonal ties that characterize the emergence and structure of networks among people such as kin or immediate work associates, and between people who hold positions in various organizations such as inter-locking directors, managers, and financial officers who associate in joint ventures (Carruthers & Babb, 2000: 49–53). Sociologists have also been concerned with specifying the benefits of various types of network relationships. For example, various types of social networks have been found to increase resources and problem-solving capabilities in business, banking, stock investing, and venture capitalization (Fernandez & Weinberg, 1997; Uzzi, 1996).

These basic concerns with enduring interpersonal ties and their advantages are relevant for considering the emergence, structure, and benefits of the increasingly extensive ties among physicians and NPCs who together deliver lasting health care services to patients in many different settings (Cooper et al., 1998; Hafferty

& Light, 1995; Sofaer, 1998). Such interrelationships now are common both in managed care organizations and in independent practice associations (Hafferty & Light, 1995; Landon et al., 1998; Laumann & Marsden, 1982; Sofaer, 1998; Stuart, 1994). For example, diabetic clinics now include NPCs (Stuart, 1994) and research shows that using NPCs and physicians in a single clinical setting increases treatment efficiency and patient satisfaction (Landon et al., 1998).

One basic principle of social networks is the idea that the *global structure* of interrelationships among participants in a network will enable and/or constrain the experiences of various participants (Cook & Whitmeyer, 1992; Wasserman & Faust, 1994). Such global forces now structuring healthcare include the multiple illnesses of many patients and their treatment needs (Aneschensel et al., 1991) such as those involving physical and mental illness (Thoits, 1994, 1995) as well as managerial mandates controlling treatment procedures and costs in many managed care organizations (Landon et al., 1998). Neither patient-provider interaction nor patterns of dyadic referral are structurally broad enough to encompass how most patients experience the totality of care they received. Moreover, while organizational theories are relevant for studying delivery options (Flood & Fennell, 1995), bureaucratic concepts are too broad to use in specifying how enduring clusters of providers and types of patients interrelate. While we apply network ideas in this paper to health care delivery, network concepts have also recently been applied to explanations of health status and health maintenance (Berkman et al., 2000). Here social networks and their global structures are used as the new units of analysis that are now being called for in order to study the logical groupings of providers structurally linked to specific types of patients (Landon et al., 1998). For example, cost efficiency is better calculated in terms of global chronic needs and considering all of the patients' illnesses would better make decisions about appropriate treatments such as length of hospital stays. Variations in the global structure of social networks of health care providers also are very relevant for understanding the outcomes of countervailing influences that are simultaneously increasing the treatment roles of NPCs and decreasing the professional dominance of physicians such as the increasing roles of nurse practitioners in physicians' offices (Chumbler et al., 2000), and the increasing use of NPCs by patients (Cooper et al., 1998; Motwani et al., 1996; Scheffler et al., 1996).

From the viewpoint of particular position-holders in a social network, the global structures of networks lead to the principle of *structural equivalence* (Burt, 1982; Cook & Whitmeyer, 1992; Galaskiewicz & Wasserman, 1993). This term refers to the roles of particular network participants – such as physicians and NPCs – and the extent to which they are delimited by the global forces clustering their actions rather than by fixed roles or dyadic exchanges among them. For example, the role of primary care physicians and specialists are increasingly defined by whether or not their collective participation in managed care arrangements is mandated by patients first seeing primary care physicians as gatekeepers. Conversely, the expanded roles of specialists and NPCs are increased when managed care arrangements allow specialists to provide treatments (e.g. vaccinations) and NPCs to diagnose and treat common illness (Cooper et al., 1998; Rosenblatt et al., 1998). Mandates that efficiently use specialists for the recurring needs of older adult patients make particular sense (Sofaer, 1998). How to efficiently structure inter-provider ties for the totality of patient needs should benefit greatly from using structurally mandated treatments and providers in hospitals, practice associations in clinics, and various types of managed care organizations including the operating procedures of Medicare (Abbott, 1988; Cooper et al., 1998; Hafferty & Light, 1995; Jones & Cawley, 1994; Sofaer, 1998; Stuart, 1994). Such structurally equivalent forms of delivery might allow NPCs and specialists to be used instead of having patients wait for the services of their primary care physicians or see other practitioners where primary care physicians are unavailable (as in rural areas).

Another basic principle of social networks, *structural conduciveness*, emerges in social contexts that enhance the need for participants to cooperate by sharing resources and following other collective norms in order to improve productivity and provide other benefits to those served. Moreover, the relative size of groups and the rates of compositional changes among interrelated groups are seen as the structural circumstances that affect probabilities of interpersonal relationships such as cooperation among the participants who belong to different groups (Galaskiewicz & Wasserman, 1993; McPherson & Ranger-Moore, 1995; McPherson et al., 1992).

The idea of structural conduciveness also has relevance for the productivity of large independent practice associations, various forms of managed care organizations, and the strategies of insurers in linking providers, hospitals, and pharmacies in ways that increase the probabilities and forms of cross-field interactions among health care providers (Burt, 1987; Galeskiewicz & Wasserman, 1993; Hafferty & Light, 1995; Kralewski et al., 1998; Landon et al., 1998; Miller, 1998; Sofaer, 1998; Stuart, 1994). For example, the relative numbers of physicians, NPCs, and alternative providers included and regularly available to be seen by patients can be better managed. The enduring global needs of patients will be better served by increasing the availability of distinct combinations of providers who are helping the patient. Physicians and chiro-practors, for example, can regularly treat chronic pain in ways that increase the shared information about pharmaceuticals, rehabilitative treatments and patient complaints (Sofaer, 1998; Taylor & Lessin, 1996). Operating principles of some hospitals, that involves more cross-department cooperation, coordination, and management of patient records by case managers involved in *Total Quality Management* (TQM), can increase structural conduciveness within organizations (Miller, 1998; Sofaer, 1998). These types of structural forces are efficiently increasing the interrelationships among diverse provider groups and are among the so-called countervailing influences appropriately reducing the hegemony of physicians (Hafferty & Light, 1995).

Another general principle of social networks is the idea of the *duality of linkages*. Network interrelationships are dualistic in the sense that positive exchanges between some dyads in a chain or in other network configurations indirectly affect other participants in that network. For example, indirect ties spread positive and cooperative relationships in chains as *A* gives to *B*, *B* gives to *C*, *C* gives to *D*, and so on. In more complex networks, trust and cooperative norms can spread through larger and larger clusters of participants by means of *bridges* (see structural holes and sub-groups below) (Burt, 1992; Cook & Whitmeyer, 1992; Uehara, 1990). Duality includes the simultaneous effects of positive exchanges among some dyads and the spread of increasingly positive dispositions among all the participants in a network, even those with distant and indirect links. Previous research has shown that physicians and NPCs whose services are linked through sharing patients have more positive views of their counterparts' professional contributions as well as becoming closer interpersonally (Chumbler & Grimm, 1994, 1995). The concept of duality of linkages is similar to the idea that experiencing interpersonal relationships of cooperation and work can reduce coworkers' sexism and racism (Taylor, 1995).

The duality of positive connectedness is particularly applicable to the patterns by which patients increasingly are referred sequentially along a chained series of providers, all of whom make positive contributions to patients' well being (Chumbler & Grimm, 1995; Grimm & Chumbler, 1995; Jones & Cawley, 1994; Motwani et al., 1996; Taylor & Lessin, 1996). Such chains increasingly exist in hospital settings that have implemented TQM (Motwani et al., 1996), in independent practice associations treating patients with chronic illnesses such as diabetes (Stuart, 1994), and because patients themselves increasingly choose alternative providers such as chiropractors as supplemental sources of treatment for pain (Eisenberg et al., 1998). What is important about all these tenets is that they provide opportunities to enhance cooperativeness and to decrease defensiveness through mutual interaction. Other examples of dualistic positive connectedness will be discussed below.

General principles of social networks are suitable for use by those who are managing comprehensive care for the general needs of patient groups or the combined needs of particular types of patients as well as by practitioners who have interest in cooperating in more comprehensive patterns of health care delivery (Cooper et al., 1998; Hafferty & Light, 1995; Kralewski et al., 1998; Landon et al., 1998; Miller, 1998). We turn now to more specific principles of social networks that provide useful insights into the nature of divergent forms of interconnectedness. In applying these principles to interrelationships among health care providers, we will be considering forms of structuring that can effectively organize more comprehensive and more effective health care.

Specific Principles

Structural encapsulation refers to the interrelated exchanges among a subset rather than among all the participants in a social network (Campbell & Lee, 1992; Uehara, 1990). In health care delivery this concept translates into the effective interconnectedness of subsets of providers relevant for particular types of illnesses or useful in certain but not all contexts of treatment such as outpatient services and home health care. For example, depending on ambulatory status, effective subsets of providers (home health care nurses, podiatrists, nurse practitioners, endocrinologists, and gynecologists) that are linked directly or indirectly could provide comprehensive annual treatment to older women (Cooper et al., 1998; Kralewski et al., 1998; Miller, 1998; Motwani et al., 1996; Scheffler et al., 1996). As the global needs of patients become the basis for structuring patients' treatment with combinations of physicians and NPCs, it is clear that the continuing set of providers can and should communicate closely and be willing to have flexible roles that ensure all needs are addressed rapidly. For example, physical therapists could be the first to recognize and communicate success of failure of pharmaceutical treatment for strokes (Ritchey, Pinkston, Goldbaum & Heerton, 1989). Among older adult patients, regular preventive health evaluations by optometrists and podiatrists could help in detecting glaucoma and diabetes, respectively (Cooper et al., 1998; Stuart, 1994). Hospitalized patients should be released by decisions based on the availability of ancillary providers such as home health nurses who can and should be more responsible for evaluating recovery and deciding on readmissions (Shi, 1996).

    Structural excludability in network involvement is determined by the boundaries of encapsulation, and restricts exchange processes to encapsulated subgroups and prevents linkages to other participants. The judgments to deny linkages to other subgroups and participants override the motives of excluded participants, who themselves may desire inclusion (Markovsky et al., 1988). In clear applicability to cost-effective clustering of health care providers, excludability would continue to be based on the traditional scrutiny and monitoring that would prevent fraud or unnecessary treatment (Jasilow et al.,

1993). In its general applicability to health care provision, structural exclud-ability would continue to involve judgments that certain providers are not necessary for certain illnesses or certain patients' treatment protocols. However, in order to apply network excludability/encapsulation, services of NPCs such as nurse practitioners and physician assistants and auxiliary groups such as physical therapists must be considered (Halpern, 1992; Jones & Cawley, 1994).

It is also important to consider the entire range of patients' needs and illnesses when judging the cost effective roles of networks of providers or by fixed roles of practitioners (Aneschensel et al., 1991). Inclusion in the integrated network and exclusion from it will not be judged on the basis of arbitrary criteria such as the professional dominance of physicians (Chumbler & Grimm, 1995; Hafferty & Light, 1995). A cost-effective approach to post-hospital care also should involve excluding services of providers that could be completed by family members, brief stays in nursing homes, or volunteers in hospices and other facilities (Gordon & Rosenthal, 1995).

Centrality and Integration

Centrality is the idea that access to and the distribution of the resources in a network occur through a single port of entry that restricts inflow and the internal dispersal of resources to other network participants (see Network 1 in Fig. 1). Consequently, the non-central participants in the network are linked only indirectly or provisionally through the central position (Cook & Emerson, 1978; Cook & Whitmeyer, 1992; Galaskiewicz & Wasserman, 1993). The essential operating feature of centralized networks is the need for controlling network resources and information on the basis of a single avenue of access. Centrality is essentially coercive and the left panel of Fig. 1 shows how this network feature could be used to restrict unnecessary health care services as well as to make available services contingent upon closer scrutiny. The relevance of the centrality principle in health care is the continuing debate about the role of "gatekeepers," such as the primary care physicians in many managed care organizations (Kralewski et al., 1998) and by the intransitive (one-directional) referral patterns from primary care physicians to specialists (Miller, 1998). *Centrality* is appropriate when scrutinizing the appropriate usage of specialists and in the cost-effective use of services such as MRI diagnosis by restricting them to a central hospital or organization (Cockerham, 1998: 259–261; Hafferty & Light, 1995). *Centrality* is much less necessary if not harmful when NPCs have expanded capabilities that allow them to play cost effective roles as first-seen providers who can make patient referrals to physicians (Hafferty & Light, 1995; Jones & Cawley, 1994; Scheffler et al., 1996; Stuart, 1994; Sofaer, 1998).

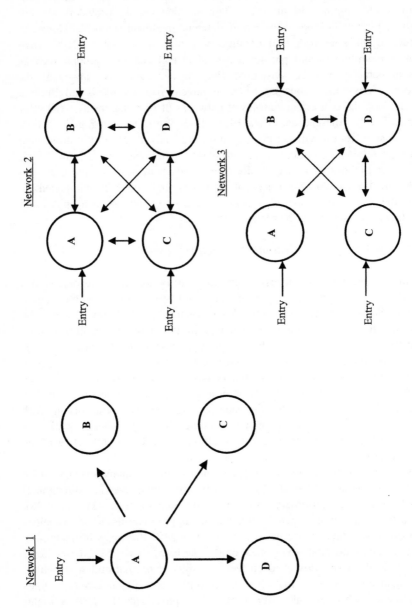

Fig. 1. A centralized network with restricted access (Network 1), an Integrated network with direct and transitive ties (Network 2), and a Semi-Integrated Network (Network 3).

The most cost-effective integration of the centrality principle can be implemented, as an organizational feature of health care, without understanding potential dysfunctions for patient needs. The most cost-effective composite treatment plans must include the recurring services of the range of providers necessary to treat all patient needs and to prevent all illnesses from getting worse (Aneschensel et al., 1991). While well intentioned, centralized medical judgments and decisions may overlook preventive and diagnostic steps that will result in much higher costs in the future (Stuart, 1994). Such a consequence is more likely when single-deciders such as primary care physicians are the only ones considering the complaints of patients. On the other hand, endocrinologists may be better able than primary care physicians to discover drug interaction effects among women taking hormonal agents in combination with other medications.

Integration occurs in social networks when there are multiple points of entry and more direct and transitive (two-way) transfers of information and resources among network positions (See Network 2 in Fig. 1) (Burt, 1987; Campbell & Lee, 1992; Carruthers & Babb, 2000). Such more flexible aspects of network structures would allow patients to be served by NPCs who may be convenient to patients as first seen providers, especially where physicians are less available (rural areas) or less cost effective treatment sources (Cooper et al., 1998). The structural flexibility illustrated in the middle panel of Fig. 1 (Network 2) shows how maximum treatment effectiveness could be globally structured if and when it is needed. The most cost-effective integration of providers for most patients probably would fall somewhere between the logical extremes of a completely centralized network or completely integrated network. Furthermore, the degree of centrality/integration will change as some physical health conditions worsen and others improve. For example, older female patients might best be served in any year by a partially integrated network comprised of an obstetrician/gynecologist (A), rheumatologist (B), endocrinologist (C), and primary care physician (D), where the points of entry for care could be any point and the major referral exchanges between all specialists and the primary care physician and between some but not all pairs of specialists (see Network 3).

There is considerable evidence that more information is shared and technological diffusion occurs faster in integrated networks. Cooperative strategizing and more equalitarian principles of operation are more widespread in integrated rather than in centralized networks (Burt, 1987; Galaskiewicz & Burt, 1991). The expanding ways in which NPCs and alternative providers such as chiropractors serve important patient needs sets structurally conducive circumstances for more integrated connectedness among providers (Cooper et al.,

1998; Eisenberg et al., 1998; Miller, 1998; Scheffler et al., 1996; Sofaer, 1998). Patterns to such integration will develop through the many types of NPCs who can provide first-seen services not usually provided cost effectively by physicians and by treatments that patients feel work in addition to those which they have received from physicians (Cooper et al., 1998; Hafferty & Light, 1995; Jones & Cawley, 1994; Stuart, 1994). Consumer oriented health care delivery is based on the comprehensive patient needs and integrated through the composite of providers that are needed to treated and prevent the progress of illnesses (Jones, 1997).

Careful decisions will be necessary when integrating the services of physicians and other providers. The fact that particular physician specialties, such as pulmonary medicine, rheumatology, obstetrics and gynecology, deliver a considerable amount of primary care to their patients is an example of the increased efficiency of integration (Rosenblatt et al., 1998). Such structured efficiency also could be cost effectively used with the expanding first-seen capabilities of NPCs, especially older adults receiving comprehensive primary care (Cooper et al., 1998; Sofaer, 1998). More points of first access also would help in addressing one of the major patient complaints about health care services – scheduling problems and delays in seeing any provider due to appointment problems. Given that the practitioner first seen can deal with the presenting complaint, integrating treatment networks to more routes of first entry, including physician extenders in emergency rooms would help reduce waiting time and delays in treatment (Ellis & Brandt, 1997; Miller, 1998).

Subgroups and Structural Holes

The dynamics of managing the changing interrelations among providers will involve the social network principles called *subgroups* and *structural holes*. *Subgroups* are the integrated clusters of network positions that are not necessarily linked continuously or directly to other integrated groupings of positions. The larger needs of managing the indirect and intermittent exchange among network sub-groups have received considerable attention from sociologists. Research has demonstrated that indirect and intermittent linkages connect organizations much the way pieces of a quilt can be stitched together so that narrow quilt pieces link larger design clusters (McPherson & Ranger-Moore, 1991; McPherson et al., 1992). Other research has focused on the changing interpersonal linkages among subgroups of neighbors, subgroups intermittently forming linkages in an organization, and people who have indirect, or intermittent linkage with various voluntary groups (Carruthers & Babb, 2000).

Translated into the context of indirect and intermittent interrelationships among health care providers, the idea of indirectly linked sub-groups provides an important mechanism for controlling the duration and relevance of providers' services. Either in the course of an illness or among different illnesses, comprehensive yet cost effective health care will involve periodic use of clusters of providers depending on patients' gender and age. Examples would be pediatricians, allergists, and pediatric surgeons for children; endocrinologists, orthopedic surgeons, and podiatrists for diabetics; and nurse practitioners, rheumatologists, and gynecologists for older women (Cooper et al., 1998; Sofaer, 1998). Subgroups of providers will vary across the onset and duration of illnesses as well as the life-course. For example, the cost-effectiveness of treatment based on subgroups in a single clinic or hospital has been well documented, as in studies of TQM (Landon et al., 1998; Shi, 1996; Westphal et al., 1997). However, very little is known about how best to access and employ sub-groups of providers as patients cross into organizational settings outside hospitals and clinics such as hospices, rehabilitation centers, residential institutions, and mental institutions (Sofaer, 1998; Thoits, 1994, 1995). A major challenge for policy makers to consider from the networks perspective is implementing the most cost-effective inter-organizational networks. To that end, the ideas of subgroup connections and structurally equivalent connections between organizations need to be implemented (Galaskiewicz & Burt, 1991; Westphal et al., 1997).

Structural holes are the positions in networks that have the potential to bridge the indirect linkage among integrated sub-groups of network participants when they are necessary (Burt, 1992). An example would include bridging positions that link sets of organizations and those that link different types of consumer groups and voluntary associations (Galaskiewicz, 1985; Laumann & Marsden, 1982; McPherson et al., 1992; McPherson & Ranger-Moore, 1991). Among interpersonal sub-groups, intermittent bridges in a single organization or social setting are those in which periodic projects call for bridges between different subgroups and the intermittently bridged clusters of support persons made of up neighbors and kin (Biggart, 1989; Campbell & Lee, 1992; Galaskiewicz & Wasserman, 1993; Granovetter, 1973; Wasserman & Faust, 1994; Uzzi, 1996). The structured efficiency of a bridging position (x) is illustrated in top panel of Fig. 2 by the single path of access to the much greater integrated resources of subgroup B that the bridge allows. In actuality, bridging positions are more likely to efficiently link partially integrated and more centralized subgroups making the importance of the bridge even greater. For example, in the bottom panel of Fig. 2, we portray how a case manager in a hospital (x) could link an elderly female patient's treatment subgroup in a multi-specialty clinic,

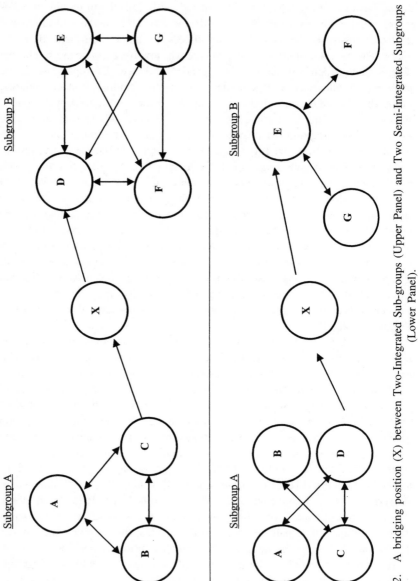

Fig. 2. A bridging position (X) between Two-Integrated Sub-groups (Upper Panel) and Two Semi-Integrated Subgroups (Lower Panel).

comprised of a gynecologist (A), rheumatologist (B), endocrinologist (C), primary care physician (D). This same case manager could also link this same older woman with a peripheral vascular surgeon (E), anesthesiologist (F) and rehabilitation specialist (G) in a hospital.

Structural holes can be viewed as bridging positions in the context of the health care delivery system. In a seminal study, Westphal et al. (1997) demonstrated that patterns of innovation spread through means other than the direct interconnections among hospitals involving interlocking directors. Instead, diffusion often occurred through the bridging connections provided by common membership in strategic alliances created by intermittently constructed agreements and by common membership in chains of proprietary hospitals. Such indirect bridging links have been demonstrated to diffuse knowledge and technology among disparate and dispersed professionals (virtual forms of bridging are discussed below) (Sproull & Kiesler, 1991). Managers of hospital chains, managers of managed care organizations, and physicians who manage large independent practice associations illustrate specific health care personnel who could serve in bridging positions (Hafferty & Light, 1995). Within hospitals and other large health care organizations, bridging positions also can be occupied by case managers and other patient-needs specialists who help patients traverse clusters of providers in different departments (Cockerham, 1998: 244–248). The coordination of treatment in hospitals would be enhanced by the use of bridging positions that intermittently link subgroups of providers. Such positions would also make patients' experiences with complex health care organizations (e.g. managed care organizations) more personable and understandable (Cockerham, 1998: 244–248).

Strong Ties and Weak Ties

One of the most insightful themes in the sociological study of social networks is the demonstrable importance of both strong ties and weak ties (Granovetter, 1973; 1983; 1985). Research has shown that both types of ties have distinct advantages. *Strong ties* are embedded in social relationships that develop through close friendships, immediate kinship, and close association in work activity (Carruthers & Babb, 2000: 49–51). *Weak ties* are relationships that are less embedded and often depend upon very indirect linkages among people or organizations. *Weak ties* are illustrated by the ties among more distantly related members of the same firm, more distant relatives, and mere acquaintances, or those who have nominal affiliations in very large voluntary associations (Carruthers & Babb, 2000; Granovetter, 1973; Uzzi, 1996).

Strong ties in work-related activity emerge from weaker ties in two types of employment relationships. The first example involves filmmaking and garment manufacturing sectors, which as individual groups have intermittent but repeated alliances among the specialized people needed to produce and market a film or type of garment (Faulkner & Anderson, 1987; Uzzi, 1996). The second is illustrated by longer-term yet variable alliances in venture capitalization and in investment banking (Baker, 1990; Fernandez & Weinberg, 1997). The emergence of stronger ties among weaker ones allows the trust and sharing of resources necessary to keep large-scale coordinated activity task-directed and yet flexible enough to allow new affiliations as they are called for when projects change. Stronger ties among health care providers emerge from among weaker associations as patient needs become clustered in complex patterns of treatment, many of which are needed for chronic diseases (Hafferty & Light, 1995; Motwani et al., 1996; Sofaer, 1998). Previous research on dyadic patient referrals shows that higher levels of trust, understanding, and respect among physicians and NPCs and physical therapists emerge through sharing patients (Chumbler & Grimm, 1995; Ritchey et al., 1989).

Considerable attention has also been paid to the easier access to information and the more rapid spread of new ideas and technology among more dispersed weak ties networks (Carruthers & Babb, 2000; Granovetter, 1973, 1983, 1985; Galaskiewicz & Wasserman, 1993). One major reason for the more extensive flow of ideas and information in weak ties networks is the greater abundance of bridges that indirectly link social circles in dispersed locations (see, especially, Granovetter, 1983). It is the weaker ties among organizations that allow information on business strategies and investment options to spread through nation-wide networks of the managerial elite (Granovetter, 1983). Bridges of weak ties linking dispersed social circles also help explain inter-personal contacts that transverse voluntary organizations (McPherson et al., 1992).

Weak ties bridges increasingly link health care organizations through such interconnectedness as chains of proprietary hospitals, contractually linked independent practice associations and various managed care organizations (Miller, 1998; Sofaer, 1998; Westphal et al., 1997). Some research argues that inter-connectedness of weaker ties explain how managerial strategies diffuse among hospitals (Westphal et al., 1997) and how routinization of treatment protocols and cost-containment strategies spread through independent practice associations and managed care organizations (Hafferty & Light, 1995). The extensiveness of inter-organizational weak ties explains the more rapid spread of certain managed care innovations in large provider networks like Kaiser Permanente (Landon et al., 1998). Weak ties arrangements in many managed care organizations similarly have the capacity to circulate procedures among physicians, NPCs, and alternative providers (e.g. chiropractors) (Landon et al., 1998).

Real Ties and Virtual Ties

Computer technology has created another realm in which the stronger and weaker ties among professionals including health care providers can exist and develop. Computer-supported social networks (CSSNs) are defined as computer networks that link people as well as machines (Wellman et al., 1996). Videoconferencing systems now allow coworkers and others engaged in the same project activity to form and sustain the multifaceted exchanges that are inherent in closer ties (Wellman et al., 1996). By their very natures, intranets within organizations allow health care providers the immediacy of direct consultation without the inconveniences of phone-calls or letters. CSSNs with visual capabilities also allow providers to interact directly with patients in completing follow-ups, screenings, and therapy sessions, especially in hard to serve rural and frontier areas of the United States (Wellman et al., 1996). Therefore, virtual ties are not necessarily dependent on the close physical proximity that typically bridges diverse health care professionals who work in the same organization (Abbott, 1988).

The role of CSSNs in developing and sustaining weak ties networks is inherent in the Internet itself as well as the accessibility of web sites and chat rooms. Evidence shows that CSSNs help develop reciprocal support, disperse information, and allow mutual aid norms to spread widely (Constant et al., 1994). CSSNs provide easy means by which large numbers of passive observers see the benefits of information sharing and other forms of cooperation. CSSNs therefore also are characterized by the duality of linkage that is manifested in real networks (Uehara, 1990). Research also shows that when organizations encourage the use of CSSNs, there is an increase in the number of weak ties their members have that transverse status levels, department lines, and other administrative barriers to information sharing (Hinds & Kiesler, 1995). In these ways, CSSNs create associational linkages that transcend organizational boundaries and divisions and allow the formation of so-called invisible colleges of dispersed colleagues (Carley & Wendt, 1991). Thus, the CSSNs could overcome some traditional organizational and locational barriers to communication among health care groups (Chumbler & Grimm, 2000). CSSNs also bring a direct and open accessibility that mutes traditional barriers of status, authority, and the sovereignty of various fields techniques and languages (Carley & Wendt, 1991). Consequently, CSSNs provide another form of communication by which traditional struggles involving the control of treatment delivery and conflicts over trust can be reduced (Ritchey et al., 1989).

Social Networks and Health Care Delivery Issues

The countervailing forces of managed care organizations, large health care contractors, and government regulations have displaced the professional dominance of physicians in health care delivery (Hafferty & Light, 1995). Further, patients have become critical consumers who increasingly share financial responsibility for their own service options (Hafferty & Light, 1995; Taylor & Lessin, 1996). Other important reasons for the diminishing global authority of physicians are the expanding roles of NPCs whose services either supplement or overlap with some of the activities of physicians in much of consolidated and managed care (Cooper et al., 1998; Kralewski et al., 1998; Miller, 1998).

Despite their diminished authority in the organization, delivery, and in the pricing of health care, physicians still retain considerable control over many diagnostic and treatment procedures; even as the autonomy and clinical privileges of NPCs have expanded and organizational mandates have made them more accessible for consumers (Cooper et al., 1998; Halpern, 1992; Light, 1993; Manley, 1995). However, in this situation of potential contentiousness, there has been no large-scale battling among physicians and other providers over treatment turf. Some physicians have joined unions in response to the encroachment of their traditional hegemony by other providers, but most have not. In fact, less than one-third of physicians now are active members of the American Medical Association, a considerably diminished political force in recent years (Cockerham, 1998: 185–189). Moreover, considerable research on the interconnectedness of physicians and other providers shows very little evidence of working relationships being characterized by rancor. Increasingly, physicians themselves use nurse practitioners and physician assistants in their own offices (Cooper et al., 1998).

The emergence of the more complex and extensive social networks among providers that link them in cooperative and workable professional arrangements coincides with this less politicized recent period of reallocation of health care activity. During the 1990s, for example, most employees in the United States became enrolled in managed care organizations and most physicians became members of or became contractually linked to managed care organization provider groups. One clear effect of the failed attempt of the Clinton Administration to increase the role of the federal government's regulatory role in health care was the rapid emergence of managed care arrangements in the health care industry (Cockerham, 1998: 269–270). It is more than coincidental that such rapid reforms took place at the same time that social networks among physicians and other providers were emerging. Such interconnectedness enabled the cooperative sharing of health care delivery in hospitals, independent practice arrangements, large multi-specialty physician conglomerates, and in managed care organizations. While it might be argued that mandates and negative incentives explain such cooperation, the power still retained by physicians makes such an argument dubious. Instead, it seems clear that the positive connectedness, and the emergence of both stronger and weaker ties among providers, will continue to permit the relatively peaceful evolution of conjoint health care arrangements.

It is plausible that the further evolution of social networks among health care providers will result in working linkages among physicians and alternative providers. Structurally conducive conditions for such interconnectedness include the increasing patient demand for services of alternative providers, especially for control of pain (Eisenberg et al., 1998); and the increasing number of managed care organizations that include alternative provider options in their contracted plans (Sofaer, 1998). Benefits of including alternative providers in the social networks among providers include better serving the needs of patients, avoiding the deleterious effects of counterproductive treatments including the use of supplements, and the benefit of the duality of linkages among providers.

Increased conflict among health care providers would be expected if the pressures for interconnectedness occur with either oversupplies of providers or undersupplies of patients to be served (Cromley & Albertsen, 1993). The distribution of physicians who are primary care providers versus those who are specialists has been an increasingly important issue in considering reforms of physician training and practice location options in the U.S. (Vanselow, 1998). It will be important to recast estimations of physician needs, however, to take into account the network-based interrelationships between them and non-physicians that could spread services effectively. For example, multiple-site practices have been shown to effectively spread the services of physicians (Cromley & Albertsen, 1993) and podiatrists (Chumbler & Grimm, 1995; Grimm & Chumbler, 1995). Recast in the framework of social networks, multiple location practices involving properly balanced interrelated services of physicians and podiatrists would more effectively structure treatment for diabetes, foot surgery, and important types of annual health screenings (Chumbler & Grimm, 1995; Stuart, 1994).

Consideration of the integrated social networks among all providers based upon a wide-ranging profile of patient needs will be a keystone for effectively meeting consumer demands (Sofaer, 1998; Taylor & Lessin, 1996). Allowing patients access to a wider-range of first-seen providers who are integrated in relevant treatment sub-groups for continuing consideration of chronic needs will have many important positive consequences. Structuring and making available such social networks of providers would allow patients to avoid unnecessary delay in first-seeing a provider and allow them the more appropriate early treatment that would preclude the more expensive treatments required as illnesses progress (Sofaer, 1998; Stuart, 1994). For example, annual transitive patient exchanges among physicians, podiatrists, optometrists and dentists would be cost effective ways of providing the early detection and follow-up services required by many geriatric patients (Cooper et al., 1998; Sofaer, 1998).

Other more integrated social networks among physicians and NPCs will effectively allow treatment delivery in emergency rooms, outpatient clinics, and in hospices that help reduce the costs of longer-term hospital care (Sofaer, 1998; Stuart, 1994). Such social networks should be structured with bridging positions that link the integrated sub-groups of providers in primary care, specialty care, and long-term care. Fuller integration of networks of providers in rural areas would allow more effective utilization of physician assistants and nurse practitioners (Chumbler et al., 2000). While monitoring is used to control costs of treatments managed in a variety of managed care organizations, it is also needed for the multifaceted exchange of information on patient treatment needs and services they have received among providers who are in social networks.

Much further consideration also must be given to the specific roles that CSSNs can play in linking providers, health care managers, and patients. The benefits of CSSNs in enhancing virtual ties as well as access to information available at web sites will play important roles in providing more opportunity to exchange information and help. CSSNs could, for example, help patients with access to providers and health care managers at web sites where answers to their treatment and coverage questions might be quickly answered. Recent studies show, for example, that nearly one-fifth of insured U.S. adults and about half of the uninsured adult population in the U.S. report problems arranging for and paying for care (Donelan et al., 1996). CSSNs arranged to provide web sites for public health information would be ideally suited to distribute the types of public information that would aid all members of the community in obtaining information about health-care and their options for obtaining it (Showstack et al., 1996).

CONCLUSION

Social networks among providers, patients, and the managers of health care will be the keystones for constructing better health care arrangements in the United States. Given the multi-payer context of care in the U.S., the various structural alternatives possible in social networks will be especially important in constructing multi-faceted combinations of care options. As these complex structures evolve and providers, managers, and patients participate in them, there is no reason to doubt that the effective and workable results observed in other industries will not develop in health care as well. Social network principles therefore hold the keys for unlocking the ways to continue the improvements in the organization and delivery of health care in the United States.

REFERENCES

Abbott, A. (1988). *The system of professions: An essay on the division of expert labor.* Chicago: University of Chicago.

Aneschensel, C. S., Rutter, C. M., & Lachenbrach, P. A. (1991). Social structure, stress, and mental health: Competing conceptual and analytical models. *American Sociological Review, 56,* 166–178.

Baker, W. E. (1990). Market networks and corporate behavior. *American Journal of Sociology, 96,* 589–625.

Begun, J. W. (1978). The consequences of professionalization for health services delivery: Evidence from optometry. *Journal of Health and Social Behavior, 20,* 376–386.

Berkman, L. F., Glass, T., Brissette, I., & Seeman, T. E. (2000). From social integration to health: Durkheim in the new millennium. *Social Science and Medicine, 51,* 843–857.

Biggart, N. W. (1989). *Charismatic capitalism: Direct selling organizations in America.* Chicago: University of Chicago.

Burt, R. (1982). *Toward a structural theory of action: Network models of social structure perception and action.* New York: Academic.

Burt, R. (1987). Social contagion and innovation: Cohesion versus structural equivalence. *American Sociological Review, 92,* 1287–1335.

Burt, R. (1992). *Structural holes: The social structure of competition.* Cambridge, MA: Harvard University.

Campbell, K., & Lee, B. (1992). Sources of personal neighbor networks: Social integration, need, or time? *Social Forces, 70,* 1077–1100.

Carley, K., & Wendt, K. (1991). Electronic mail and scientific communication. *Knowledge, 12,* 406–440.

Carruthers, B. G., & Babb, S. L. (2000). *Economy/Society: Markets, meanings, and social structure.* Thousand Oaks, CA: Pine Forge.

Chumbler, N. R., Geller, J. M., & Weier, A. W. (2000). The effects of clinical decision-making on nurse practitioners' clinical productivity. *Evaluation & the Health Professions, 23,* 284–305.

Chumbler, N. R., & Grimm, J. W. (2000). Channels of podiatrists' referral communication to physicians. *Journal of Applied Sociology, 17,* 69–85.

Chumbler, N. R., & Grimm, J. W. (1994). An explanatory model of reciprocal referrals between podiatrists and physicians. *Journal of Applied Sociology, 11,* 59–76.

Chumbler, N. R., & Grimm, J. W. (1995). Reciprocal referrals between podiatrists and physicians: The effects of professional training, practice location, and non-medical reasons for referrals, In: J. J. Kronenfeld (Ed.), *Research in the Sociology of Health Care* (Vol. 12, pp. 261–285). Greenwich, CT: JAI.

Chumbler, N. R., & Grimm, J. W. (1996). Surgical specialization in a limited health care profession: Countervailing forces shaping health care delivery. *Free Inquiry in Creative Sociology, 24,* 59–66.

Constant, D., Kiesler, S. B., & Sproull, L. S. (1994). What's mine is ours, or is it?: A study of attitudes about information sharing. *Information Systems Research, 5,* 400–421.

Cook, K. S., & Emerson, R. M. (1978). Power, equity, commitment, and exchange networks. *American Sociological Review, 43,* 721–739.

Cook, K., & Whitmeyer, J. (1992). Two approaches to social structure: Exchange theory and network analysis. *Annual Reviews of Sociology, 18,* 109–127.

Cooper, R., Henderson, T., & Dietrich, C. (1998). Roles of non-physician clinicians as autonomous providers of patient care. *Journal of the American Medical Association, 280,* 795–801.

Cromley, F. K., & Albertsen, P. C. (1993). Multiple-site physician practices and their effect on service distribution. *Health Service Research, 28,* 503–522.

Donelan, K., Blendon, R., Hill, C., Frankel, M., Hoffman, C., Rowland, D., & Altman, D. (1996). Whatever happened to the health insurance crisis in the United States? Voices from a National Survey. *Journal of the American Medical Association, 276,* 1346–1352.

Ellis, G., & Brandt, T. (1997). Use of physician extenders and fast tracks in the United States: Emergency departments. *American Journal of Emergency Medicine, 15,* 229–232.

Eisenberg, D., Davis, R., Ettner, S, Appel, S., Wilkey, S., Rompay, M., & Kessler, R. (1998). Trends in alternative medicine use in the United States: 1990–1997. *Journal of the American Medical Association, 280,* 1569–1575.

Faulkner, R. R., & Anderson, A. B. (1987). Short-term projects and emergent careers: Evidence from Hollywood. *American Journal of Sociology, 92,* 879–909.

Fernandez, R., & Weinberg, N. (1997). Sifting and sorting: Personal contacts and hiring in a retail Bank. *American Sociological Review, 62,* 883–902.

Flood, A. B., & Fennell, M. L. (1995). Through the lenses of organizational sociology: The role of organizational theory and research conceptualizing and fixing our health care system. *Journal of Health and Social Behavior, Extra Issue,* 154–169.

Galaskiewicz, J. (1985). Professional networks and the institutionalization of a single mindset. *American Sociological Review, 50,* 639–658.

Galaskiewicz, J., & Burt, R. (1991). Interorganization contagion in corporate philanthropy. *Administrative Science Quarterly, 36,* 88–105.

Galaskiewicz, J., & Shatin, D. (1981). Leadership and networking among neighborhood human service organizations. *Administrative Science Quarterly, 26,* 434–448.

Galaskiewicz, J., & Wasserman, S. (1993). Social network analysis: Concepts, methodology, and directions for the 1990s. *Sociological Methods and Research, 22,* 3–22.

Gordon, H. S., & Rosenthal, G. E. (1995). Impact of marital status on outcomes in hospitalized patients. Evidence from an academic medical center. *Archives of Internal Medicine, 155,* 2465–2471.

Granovetter, M. (1973). The strength of weak ties. *American Journal of Sociology, 78,* 1360–1380.

Granovetter, M. (1983). The strength of weak ties: A network theory revisited. In: R. Collins (Ed.), *Sociological Theory* (pp. 201–233). San Francisco: Jossey-Bass.

Granovetter, M. (1985). Economic action and social structure: The problem of embeddedness. *American Journal of Sociology, 91,* 481–510.

Grimm, J. W., & Chumbler, N. R. (1995). The role of network strength in patient referrals between podiatrists and physicians. *Sociological Imagination, 32,* 98–118.

Hafferty, F., & Light, D. (1995). Professional dynamics and the changing nature of medical work. *Journal of Health and Social Behavior, Extra Issue,* 132–153.

Halpern, S. (1992). Dynamics of professional control: Internal coalitions and cross-professional boundaries. *American Journal of Sociology, 97*, 994–1021.

Han, S. (1996). Structuring relations in on-the-job networks. *Social Networks, 18*, 47–67.

Hinds, P., & Kiesler, S. (1995). Communication across boundaries: Work, structure, and use of communication technologies in a large organization. *Organization Science, 6*, 373–393.

Jasilow, P., Pontell, H., & Geis, G. (1993). *Prescription for profit: How doctors defraud Medicaid.* Berkeley, CA: University of California Press.

Jones, K. (1997). Consumer satisfaction: A key to financial success in the managed care environment. *Journal of Health Care Finance, 23*, 21–32.

Jones, P., & Cawley, J. (1994). Physicians assistants and health system reform: Clinical capabilities, practice activities, and potential roles. *Journal of the American Medical Association, 271*, 1266–1272.

Kerssens, J., & Groenewegen, P. (1990). Referrals to physiotherapy: The relation between the number of referrals, the indication for referral, and the inclination to refer. *Social Science and Medicine, 30*, 797–804.

Kralewski, J., Rich, E., Bernhardt, T., Dowd, B., Feldman, R., & Johnson, C. (1998). The organizational structure of medical group practices in a managed care environment. *Health Care Management Review, 23*, 76–96.

Landon, B., Wilson, I., & Cleary, P. (1998). A conceptual model of the effects of health care organizations on the quality of medical care. *Journal of the American Medical Association, 279*, 1377–1382.

Laumann, E., & Marsden, P. (1982). Microstructural analysis in interorganizational systems. *Social Networks, 4*, 329–348.

Light, D. (1993). Countervailing power: The changing character of the medical profession in the United States. In: F. Hafferty & J. McKinley (Eds), *The Changing Medical Profession: An International Perspective* (pp. 69–79). New York: Oxford University Press.

Manley, J. (1995). Sex-segregated work in the system of professions: The development and stratification of nursing. *The Sociological Quarterly, 36*, 297–314.

Markovsky, B., Willer, D., & Patton, J. (1988). Power relations in exchange networks. *American Sociological Review, 53*, 220–236.

Marsden, P. (1983). Restricted access in networks and models of power. *American Journal of Sociology, 88*, 686–717.

McPherson, J. M., Popielarz, P. A., & Drobnic, S. (1992). Social networks and organizational dynamics. *American Sociological Review, 57*, 153–170.

McPherson, J. M., & Ranger-Moore, J. (1991). Evolution on a dancing landscape: Organizations and networks in dynamic Blau Space. *Social Forces, 70*, 19–42.

Miller, R. (1998). Health care organizational change: Implications for access to care and its measurement. *Health Services Research, 33*, 653–680.

Mizruchi, M. S. (1996). What do interlocks Do?: An analysis, critique, and assessment of research on interlocking directorates. *Annual Review of Sociology, 22*, 271–298.

Motwani, J., Sower, V., & Brashier, L. (1996). Implementing Total Quality Management in the health care sector. *Health Care Management Review, 21*, 73–82.

Munroe, B., & Steele, J. (1998). Foot-care awareness: Survey of persons aged 65 years and older. *Journal of the American Podiatric Medical Association, 88*, 242–248.

Ritchey, F., Pinkston, D., Goldbaum, J., & Heerton, M. (1989). Perceptual correlates of physician referral to physical therapists: Implications for role expansion. *Social Science and Medicine, 28*, 69–80.

Rosenblatt, R., Hart, G., Baldwin, L., Chan, L., & Schneeweiss, R. (1998). The generalist role of specialty physicians. *Journal of the American Medical Association, 279*, 1364–1370.

Scheffler, R., Waitzman, N., & Hillman, J. (1996). The productivity of physician assistants and nurse practitioners and health work force policy in the field of managed health care. *Journal of Allied Health, 25*, 207–217.

Shi, L. (1996). Patient and hospital characteristics associated with average length of stay (ALOS). *Health Care Management Review, 21*, 46–61.

Showstack, J., Lurie, N., Leatherman, S., Fisher, E., & Inui, T. (1996). Health of the public: The private-sector challenge. *Journal of the American Medical Association, 276*, 1071–1074.

Sofaer, S. (1998). Aging and primary care: An overview of organizational and behavioral issues in the delivery of health care services to older Americans. *Health Services Research, 33*, 298–321.

Sproull, L. S., & Kiesler, S. B. (1991). *Connections: New ways of working in the networked organization.* Boston: MIT Press.

Stuart, M. (1994). Redefining boundaries in the financing and care of diabetes: The Maryland experience. *The Milbank Quarterly, 72*, 679–694.

Taylor, M. C. (1995). White backlash to workplace affirmative action: Peril or myth? *Social Forces, 73*, 1385–1414.

Taylor, R., & Lessin, L. (1996). Restructuring the health care delivery system in the U.S. *Journal of Health Care Finance, 22*, 33–60.

Thoits, P. A. (1994). Stressors and problem solving: The individual as psychological activist. *Journal of Health and Social Behavior, 35*, 143–159.

Thoits, P. A. (1995). Stress, coping, and social support processes: Where are we? What next? *Journal of Health and Social Behavior*, (Extra Issue), 53–79.

Uehara, E. (1990). Dual exchange theory, social networks, and informal support. *American Journal of Sociology, 96*, 521–527.

Unland, J. (1998). The range of providers/Insured configurations. *Journal of Health Care Finance, 24*, 1–25.

Uzzi, B. (1996). The sources and consequences of embeddedness for economic performance of organizations: The network effect. *American Sociological Review, 61*, 674–698.

Vanselow, N. (1998). Primary care and the specialist. *Journal of the American Medical Association, 279*, 1394–1395.

Wasserman, S., & Faust, K. (1994). *Social network analysis: Methods and applications.* Cambridge, U.K.: Cambridge University Press.

Wellman, B., Salaff, J., Dimitrova, D., Garton, L., Gulia, M., & Haythornwaite, J. (1996). Computer networks as social networks: Collaborative work, telework, and virtual community. *Annual Review of Sociology, 22*, 213–238.

Westphal, J. D., Gulati, R., & Shortell, S. M. (1997). Customization or conformity? An institutional and network perspective on the content and consequences of Total Quality Management adoption. *Administrative Science Quarterly, 42*, 366–394.

PART II:
SPECIFIC TECHNOLOGIES
AND PROGRAMS

HOSPITAL TECHNOLOGY-ENVIRONMENT INTERPLAY AS DETERMINANTS OF MORTALITY

Eleanor V. Toney

ABSTRACT

This study examines healthcare from an environmental point of view using the ecology theory of organizations as the underlying framework. This research attempts to understand how hospital services and their technology affect the community. The focus is on community characteristics, hospital services and technology available within the community and their influence on community mortality rates. The community is defined as a Healthcare Service Area (HCSA) which is determined by the hospital utilization pattern of individuals. Data from the 1995 Area Resource Files are utilized in this analysis. Lisrel, structural equation modeling, was utilized for data analysis. The results indicate that socioeconomic status, presence of teaching hospitals and the age of the population will have a greater influence on crude mortality rates than the actual services and technology that hospitals provide. In summary, the findings suggest that the discussion of healthcare needs to look beyond the hospitals and their high tech services and diagnostics to determine what services will actually benefit the community.

Changing Consumers and Changing Technology in Health Care and Health Care Delivery,
Volume 19, pages 83–101.
Copyright © 2001 by Elsevier Science Ltd.
All rights of reproduction in any form reserved.
ISBN: 0-7623-0808-7

INTRODUCTION

This study looks at healthcare from an environmental point of view using the ecology theory of organizations as the underlying framework. The study focuses on community characteristics, the hospital services that are available and their influence on community mortality rates. Data from the 1995 Area Resource Files are utilized in this analysis. The focus of this research is the influence hospitals have on their environment versus the standard approach of studying the influence of environment upon a hospital. The findings of this study support the notion that hospitals compete amongst themselves to acquire expensive technology and offer high technology treatments such as intensive care units, regardless of the population needs (Fennell, 1980). Competition for status or prestige drives the hospitals to develop services that tend to be relatively expensive but have little influence on overall mortality rates.

I posit that if the community within which it is situated influences an institution, then too the community is influenced by the presence of the institution. Ideally, if the imposed cost cutting and healthcare reforms have made the hospitals more competitive and efficient, therefore enabling them to offer better care to the patients, then hospitals should be contributing to a lowering of the overall community mortality rates. Previous research has primarily focused on hospital morality rates within metropolitan areas. I argue that studying medical outcomes in only metropolitan hospitals is not an adequate strategy for understanding the relationship that exists between a hospital and its environment. By using the variables that have been considered when looking at hospital mortality and statistically determining if these same variables have any relationship to the mortality rates in the communities in which the hospitals are located, I will demonstrate that much of the technology that hospitals have acquired as well as the intensive services that they offer have had very little relationship to mortality rates.

Environmental Framework

The ecology model of organizations suggests that an organization will develop and respond to changes within its operating environment (Perrow, 1986; Hannan & Freeman, 1977). A population ecology model would identify the groups of organizations that have developed to serve the medical needs of the community's citizens. This model would take into account aspects of niche and competition for necessary resources when studying the types of healthcare organizations and the services that they offer. These organizations are interdependent and compete for dominance within the community. The forms of the healthcare organizations within the community can be simplistic such as a single community hospital or multiple and diverse such as multi hospital systems with various group or clinic practices. Perrow (1986), however, sees that there is more of a give and take when using the population ecology model to study organizations. While the environment will influence a given institution, the institution will have some impact upon the environment within which it operates.

Hannan and Freeman (1977, 935) view organizations as distinct species that develop within a particular environment. How these organizations grow or differentiate is dependent upon competing organizations and economic resources available within an environment. However, when using the biological study of ecology, the narrow focus on just the organism and its obvious surroundings does not provide for a complete understanding of an organism's ecological impact. To study just the organism and its immediate surroundings will not necessarily provide a complete picture of all the various interactions that occur in order for that organism to survive. Also, all organisms impact the environment within which they exist, consciously or unconsciously, either in a positive or negative manner with a balance being established within the ecosystem. An area or environment might have two or more hospitals but the services that they provide may serve a completely separate set of clientele, a complementary coexistence. Some hospitals may choose to focus on only one aspect of healthcare such as pediatrics or rehabilitation, while others may have a wider range of services to meet the needs of all age groups. Within a particular environment, some hospitals may attract national and/or international clients while the other hospitals cater to the local needs. While a hospital's ability to survive and grow is constrained by the economic resources of the citizens it serves, there are also larger external controlling factors such as state and federal laws that regulate services and determine the amount of payment for certain services or conditions. These larger external regulators might be viewed as climatic changes that are not truly under the control of any one group or organization. Such climatic changes could be sufficiently severe to cause a hospital or healthcare institution to close. The closure of a hospital in a small rural community may be considered catastrophic if only the rural region is the context of analysis. On a larger scale the closing of the hospital may have little impact upon the actual health and well-being of the citizens as they may disperse to other hospitals in the area.

Organizations are not passive entities within an environment, just as biological organisms are not passive within their environments. There is more than static tension between predator and prey. Perrow (1986, 188) suggests that the hospitals will attempt to control and manipulate the environment versus being passive participants. Therefore changes in economic resources, the population that a hospital serves and state and federal legislation may influence the services that a hospital offers and ultimately influence the quality of life for the population that the hospital serves. This notion of competition or survival of the fittest in healthcare is not as clear-cut as it might be for other organizations such as newspapers. The notion that more hospitals will compete to provide cost effective and efficient services to customers and thereby decrease the cost of healthcare has not been borne out. In particular, while a customer may shop for the lowest price for healthcare insurance, they will not shop around for the lowest cost for gallbladder surgery should this problem arise. Hospitals will not be offering sales on healthcare services in order to attract customers to come to their institutions. They may attempt some marketing of the quality of services that they provide and they may indeed look for ways in order to cut the overall costs for particular healthcare needs but in general the competition is much more muted. Hospitals compete for insurance dollars and not customers. To persuade insurance carriers to prefer their hospital instead of a competitor, hospitals may purchase high-tech diagnostic devices such as CT-scanners or offer the latest treatment options. Therefore there may be multiple high tech X-ray devices within a region or there may be a number of hospitals that offer high tech treatments such as open-heart surgery or angioplasty. This type of competition within an area will not necessarily mean lower priced healthcare with one hospital finally winning or driving the other hospital into extinction. In general this type of competition helps to eliminate or absorb other, more threatening sources of competition such as government managed healthcare. The public in general is served poorly when this adjustment occurs (Perrow, 1986, 188). Therefore, while there may be numerous high tech services and treatments offered by hospitals, there may not be an overall reduction in mortality rates for a community. It is the premise of this study that hospitals and hospital services that are being provided will have a positive relationship to actual well-being of the community within which they operate.

DATA AND METHODS

This study uses the community or area within which healthcare is provided, as the unit of analysis. The theoretical model for the interaction of the environment, the hospital and its technology and the outcome variables, infant mortality and mortality, is shown in Fig. 1. The environment will influence what types of technology and services a hospital will develop, which should ultimately influence the well-being of the community by being inversely related to the mortality rates for infants and older adults. Therefore, there is interplay between the hospital and the environment. The mortality rates for infants and for older adults will be a direct result from this exchange of resources.

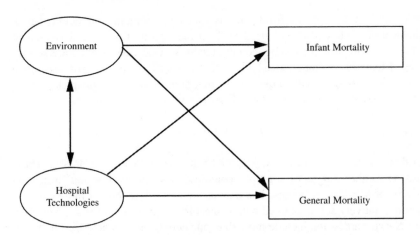

Fig. 1. Conceptual Framework with Possible Relationship for Hospital Services and the Concepts of Mortality.

Structural equation modeling will be utilized to study the relationship between hospital attributes, community attributes and mortality rates. Structural equation modeling, in particular, Lisrel allows for the analysis of constructs that do not have an agreed upon unit of measurement. For example the construct of competition is not directly observable or measurable since there is no agreed upon measure for it. The construct of competition, may be measured by proxy indicators or measurable variables, such as total number of hospital beds or total number of hospital administrators. With this modeling approach one can estimate the relationship between the observed variables and the unobserved theoretical constructs including the amount of measurement error associated with the indicators and their construct. Also structural equation model allows one to establish relationships among multiple variables at the same instance. This allows for the removal of variables that do not function as indicators for a construct as well as the removal of indicators that are closely correlated with one another. This method also allows one to validate the goodness of fit for each measurement model and for the overall structural model, providing a means to validate the appropriateness and stability of each model.

Data Source

The data set utilized are the Area Resource Files (ARF), 1995, produced by the Office of Research and Planning, Bureau of Health Professions. These files are composed of surveys from various medical, hospital, and public health organizations. They also include data from the Census Bureau and Bureau of Labor. Data are sorted by state, county, health service area, health care service areas, Medicare payment areas as well as census tracts. A cross sectional analysis of this data will allow one to determine if a relationship does exist between hospital services and mortality rates in their environment.

Unit of Analysis

For this study, the unit of analysis is the healthcare service area (HCSA). These areas are either single or multiple counties, grouped by hospital utilization patterns of Medicare residents. These areas are defined as "one or more counties that are relatively self-contained with respect to the provision of routine hospital care. Service areas that include more than one county are characterized by travel between the counties for routine hospital care" (Makuc et al., 1991, 2). While this unit of analysis was first established to look at Medicare patients and hospital utilization, it subsequently has been used to look at hospital utilization by pregnant Medicaid patients (NCHS-National Maternal and Infant Health Survey, 1991). This then represents the geographical area in which hospitals will compete for, receive and treat patients.

Using HCSA as the unit of analysis provides a different perspective on the type of clients a hospital is treating. Clients within a HCSA may reside in rural communities or perhaps even at the border between states. The location of the client in relation to where the client drives to obtain medical care will determine the amount of healthcare resources that are available to these clients and ultimately impact upon the health of this person requiring hospitalization. The patients may live in neighboring counties, or, if the hospital is close to a state border, the patients may come from another state. Thus the hospital located in an urban area may still be providing care to patients that are living in rural counties. Where a patient lives will influence the amount of preventative medicine they may have received, the illnesses that they are most likely to have experienced and the severity of the condition at the time the patient arrives at the hospital. By using the healthcare service areas as the unit of analysis, a broader, more detailed, picture is obtained for the type of patient a hospital is likely to treat and the resources that are available to the client and the hospital. Since the healthcare service area is based on the distance that Medicare patients travel to receive care, it provides a precise geographically defined area. These defined geographical regions become the environments within which the hospital operates. I hypothesize that if the hospital has sufficient technology and services, then the overall mortality rate for the service area should decrease. A negative relationship should be functioning between HCSA mortality rates and hospital characteristics. The more hospitals and hospital services available, the lower the mortality rate within a HCSA.

There are 816 healthcare service codes identified by the National Center for Health Statistics. Since areas such as the State of Alaska, the Hawaiian Islands and Honolulu did not have healthcare service area codes initially, they are provided numbers that followed in sequence to the 816 original codes. The ARF identifies 807 healthcare service areas including Alaska and the Hawaiian Islands.[1]

Constructs

Six constructs were formulated to model the relationships between the healthcare service area demographics: economic resources, competition within the region amongst the various providers, the influence of teaching institutions, the high tech treatment options provided by hospitals, diagnostic abilities and number of physicians within a service area. These constructs are labeled and represented by the latent variables; economic, compete, teach, RX, DX, and MDs. See Table 1 for the definition of the respective constructs and their indicators.

Table 1. Constructs with their Corresponding Observable Measures.

Environment	**Institutional Technologies**
Economics (ECONOMIC)	Treatments (RX)
Number below	Open-heart surgery (OHPOP)
poverty level (POVPOP)	Angioplasty (ANGIPOP)
	Cardiac Catheterization Labs (CCLAPOP)
Physicians(MDs)	**Diagnostics (DX)**
Board Certified MD's (BDCTPOP)	CT-Scanners (CTSCPOP)
	MRI Scanners (MRIPOP)
	Orthopedic Surgery (ORTHPOP)
	Joint Commission on Accreditation
Elderly	(JCAHPOP)
Number of individuals between	
65 and 84 (TOTELD)	
Competition (COMPETE)	**Teaching (TEACH)**
Full-time Administrators (ADMPOP)	Number on HMO Members (HMOMPOP)
Total number of Hospital Beds	Total Number of Teaching Hospitals
(TOBEPOP)	(TTHOPOP)

The number of individuals who are below the poverty level measures the construct of economics. Measures such as per capita income, high school education level and amount of unemployment have all been utilized as a measure for determining the socioeconomic well-being of a region. All of these variables are highly interrelated and therefore possessed collinearity that was too high to be useful in this model. However, measures such as per capita income or education may not adequately demonstrate the socioeconomic status of a region with respect to health care utilization. Using the number of individuals living below the poverty level for a HCSA attempts to measure those individuals that are at the greatest risk of lacking health insurance coverage, or preventative care. Therefore, using the poverty level will be a more inclusive measure of community well-being versus using a specific race/ethnicity composition or percent of children under the age of 18.

While the inclusion of the number of teaching hospitals under the construct of teaching is fairly obvious, the inclusion of the number of members within health maintenance organizations is less obvious. Thorpe (1997) has noted that as cost-containment is implemented, teaching hospitals, which have higher costs associated with their teaching mission as well as a larger volume of uncompensated care, have a much harder time budgeting their revenues and resources. HMOs exert a similar effect upon teaching hospitals, since they negotiate payment rates with the individual hospitals. Teaching hospitals become limited in their ability to shift the costs to maintain their teaching mission and provide indigent care to private insurance payers. Areas that have a large number of health maintenance organization members will have fewer teaching hospitals or teaching hospitals with limited services and severe financial difficulties.

Outcome Measures

The outcome measures utilized in this study are infant mortality and overall mortality rates. Infant mortality was utilized since it is highly associated with socioeconomic conditions and areas that have poor prenatal care. Though there is some overlap between this construct and the second construct of generalized mortality, this construct separates the health care for individuals of childbearing age from those who are elderly. Hospitals that provide services for infants, such as neonatal intensive care units, should decrease the occurrences of infants dying prior to reaching one year of age. This construct is measured by using the 3-year average for infant mortality for the counties which reside within the HCSA. Since infant mortality is not an extremely common occurrence and due to restrictions placed by the National Center for Health Statistics for releasing county-level information in which there are fewer than three occurrences, a 3-year averaged infant mortality rate is utilized.

The second outcome measure utilized is the general mortality rate. This mortality rate is obtained by measuring the mortality rate due to various diseases. This construct uses the data from mortality due to the most common disease processes: cancer, ischemic heart disease, flu and pneumonia and deaths due to other cardiovascular causes. While there is some collinearity among these diagnoses, they do provide a means to look at common causes of mortality.[2]

Variables

Because the HCSA level computations produce variables that are highly interrelated and possess a high degree of multicollinearity, each variable was divided by its population. This per capita standardization assisted in removing some of the scale factor that was present due to the variation in the population sizes of the healthcare service areas. Indicators such as number of intensive care units, open heart surgical units and CT and MRI scanners represent technology present within HCSAs to treat or diagnosis illnesses. To measure competition, variables such as the number of joint commission accredited hospitals (JCAH), administrators, health maintenance organizations (HMOs) and hospital beds within an area are used.

Results

Correlation statistics for the variables show that some of the variables are closely related to one another. For example, the variable, open-heart surgical units, is also related to angioplasty. However, these two variables load high for the latent variable of treatment at 0.88 for open-heart units and 0.79 for angioplasty units with only a small error value. This suggests that these treatments, even though they are interrelated, do provide a means to objectively measure high tech treatment options available within healthcare service areas.

Measurement Model

The measurement model for the proposed indicators and latent variables is presented in the form of a confirmatory factor model in Fig. 2.[3] The results show that the indicators for the proposed latent variables are highly weighted on the latent variables, suggesting that they are appropriate observable indicators for those constructs. The Chi-square of 184.41 with 52 degrees of freedom (df) is significant, and the overall goodness of fit (GOF) for this model of 0.97

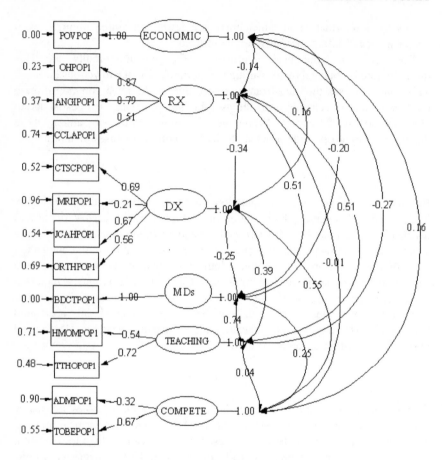

Chi-Square=185.41, df=52, P-value=0.00000, RMSEA=0.056

Fig. 2. Measurement Model.

suggests that the indicators and constructs provide an adequate representation of the empirical data.

The correlations, (ϕ), between the latent variables, treatments, diagnostics, MDs, economic, and teaching shows the direction and strength of the relationship between these variables. The latent variable, diagnose, is negatively related to treatment which one would expect. The better the diagnostic ability, the less need for expensive treatments. The correlation is -0.34 for diagnostics and treatment. The latent variable MDs is positively related to treatments but negatively associated with diagnosing. The relationship between MDs and treatment is strong at 0.51 but is weaker with a −0.25 for the relationship to the construct diagnoses. Again this is supported in reality since MDs control the treatment options for clients, but they generally do not control the presence of diagnostic equipment. Therefore, with more physicians within an area there might be a decreased need for expensive diagnostic equipment. The only correlations that are quite small and statistically not significant are between the latent variable competition and treatment with a −0.01 value and with competition and teaching with a 0.04 value. The indicators utilized to measure competition, the total number of hospital beds (TOBEPOP) within an area loading at 0.67 and the number of full-time administrators (ADMPOP) loading at 0.32 suggests a strong relationship. Therefore it seems that the latent variable, competition, while it has a strong relationship with the latent variable diagnostic at 0.55, MDs at 0.25 and economics of the area with 0.16, has little influence on the treatments available or whether there are teaching hospitals within the area.

Structural Models

Three structural equation models were developed to study the relationship between environment, hospital services and mortality. The first structural model looked at infant mortality as the dependent variable. This model utilized the six constructs to measure the interactions between healthcare and environment. The second structural model looked at general mortality issues. It includes the seventh construct measured by the number of elderly individuals in a healthcare service area. The third model combines the first two models into one.

The first structural model shows the relationship between the constructs for the community, the hospital and the dependent variable, infant mortality. This model has a Chi-square of 204.25 with 59 df, $p = 0.000$ and a GOF of 0.97. It appears to be a valid model for explaining infant mortality in terms of the identified constructs.

From Fig. 3 you can see that poverty level seems to have the greatest influence upon infant mortality. Poverty has an effect parameter of 0.45 and is the highest loading factor for this model. The construct of treatments has very little influence on infant mortality with a loading factor of 0.01. Since these infants are generally of low birth weight, the treatments available within a healthcare area would not be beneficial. Items such as prenatal care or preventative treatments are more important than the post delivery treatments, which are after the fact. This is supported by the CDC reports for the National Maternal and Infant Health Survey (NMIHS). The next construct that seemed to have a positive impact upon infant mortality was the construct of teaching. Teaching had an

94

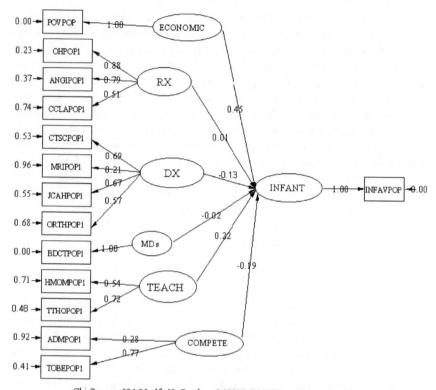

Chi-Square=204.25, df=59, P-value=0.00000, RMSEA=0.055

Fig. 3. Standardized Coefficients for Infant Mortality Structural Equation.

effect of 0.22 on infant mortality. This suggests that the more teaching facilities and the more members of HMOs present within an area, the greater the likelihood for infant mortality. While this finding for the construct of teaching was not supported in a study by Horbar et al. (1997), it was supported by Pollack et al. (1994). Studies that looked at the general population and teaching hospitals have found that mortality rates were lower in these hospitals (Shortell & Hughes, 1988; Hartz et al., 1989). This could be due to less of an emphasis on preventative prenatal care and more emphasis placed on high tech treatments. This finding could be demonstrating the negative effects that HMO membership and other cost containment policies have had upon the financial health of the teaching hospital. The teaching hospitals may have closed the prenatal clinics since they tend to not generate the revenue that neonatal intensive care units generate as one possible means to maintain financial stability. The burden of uncompensated care could be stretching the financial resources of the teaching hospital so that it has redirected its energies away from preventative health to providing care only to those who require intensive treatment. However, this redirection of services may mean those who are being cared for in these units are sicker and much more likely to die. This result requires cautious interpretation since teaching hospitals differ dramatically even among themselves. The supervision of the residents, the number of available specialists and sub specialists and the staffing patterns for medical students, interns and residents is not measured in this construct. All of these factors could be present and therefore influencing patient outcomes.

Two constructs had an inverse relationship to infant mortality. Competition and diagnostics have a mild negative relationship with infant mortality of -0.13 and -0.19 respectively. Only the variable competition is statistically significant for this model. This suggests that with more competition there is less infant mortality. This could be due to a generalized improved economic situation within the service area that allows for competition to develop and clients with sufficient economic resources to afford the diagnostic and prenatal care that would improve the outcome for the infant. In economically unstable areas there are fewer physicians, less competition and less diagnostic services, which means less prenatal and post natal care available for the clients resulting in a higher infant mortality. In particular this construct demonstrates the interplay between environment, hospital services and the outcome measure of infant mortality.

The second structural model looks at mortality in a more generalized fashion. This model includes the construct of elderly, since they tend to utilize a large amount of healthcare services, spend the most in healthcare dollars and are more likely to die within a community. This model is also valid in that although it has a Chi-square of 501.66 and 110 dfs and $p = 0.000$, the GOF index is 0.94. Figure 4 shows the relationship between the identified constructs and mortality. The main differences that are obvious from the previous model are in the constructs of poverty and elderly. Poverty in the second model shows a decrease in its effect to 0.12 while the parameter effect for elderly is 0.95. This suggests that the elderly individuals in the area have other resources that allow them to take advantage of healthcare services, though those residing in poorer areas still face higher net mortality rates. The effects of poverty, though still present, are not as great for those who have attained some age as it is for the infant and the very young. The more elderly members present within a community increases the likelihood of death by the very virtue that they are aging. The construct of treatment, changes from having a minimally positive relationship (0.01) to infant mortality to a small negative association (-0.03) when studied in the context of general mortality, though it is not statistically

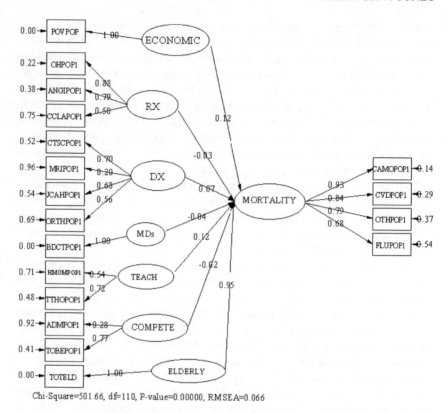

Fig. 4. Standarized Coefficients for General Mortality Structural Equation Model.

significant. This suggests that having treatment options available will at least help some older individuals. Having technologically advanced treatment options available for infants may only prolong an inevitable death, whereas for an older individual it may actually prevent the death from occurring. Since treatment can only occur after a diagnosis, the shift from a statistically non-significant, negative relationship between diagnostics and infant mortality in the first model, to a statistically significant positive relationship for general mortality in the second model is rather important. For infants the ability to provide accurate diagnostics allowed one to prevent a possible death, but for the older individual, this same ability is now positively associated with mortality. This suggests that hospitals with more diagnostic services may find more illnesses in the older patient that are not amenable to treatment, such as certain forms of cancer, and therefore result in a positive association with mortality rates. However, being able to provide diagnostic services for infants would allow for treatments that could be lifesaving and thereby decrease infant mortality.

The number of physicians within an area continues to influence the care and subsequently decrease the mortality rate within a HCSA as is evident with the slightly larger inverse relationship to general mortality, though this construct remains statistically non-significant. The construct of teaching has a positive association with mortality and this supports the studies by Shortell and Hughes (1988), Hartz et al. (1989), Kuhn et al. (1994). While the relationship between teaching hospitals and general mortality remains positive, it exerts much less of an effect upon general mortality than it did upon infant mortality. The financial problems that teaching hospitals are currently experiencing may provide a partial explanation for the positive association to mortality. However, this finding also suggests that perhaps older individuals are more resilient and can better tolerate the teaching hospital environment. Competition on the other hand became less of a factor for general mortality rates than it was for infant deaths. There are several plausible explanations for why competition is no longer significant in reducing general mortality, especially the possibility that competition is not being adequately measured. If in fact the construct of competition is being appropriately measured, then this construct demonstrates that the drive to have the latest high tech services is not aimed at serving the individual.

The third structural model is a combination of the first two models, including both infant and general mortality outcomes. Figure 5 shows this model and the resultant relationships. The Chi-square is 543.50 with 121 dfs and $p = 0.0000$. The goodness of fit (GOF) for this model has decreased from the previous two models but is still acceptable at 0.93. The factor loading for the observable measures are essentially the same as with the previous two models. The socioeconomic status of the patient continues to exert a positive influence upon infant and general mortality rates. For infants, the influence of socioeconomic status remains strong suggesting that the infant's health is highly influenced by the ability to afford prenatal healthcare services. Treatment is the only variable that changed in the direction of its influence upon infant mortality when placed in a model that considers both infant mortality and general mortality, though it remains non significant statistically. In this model, treatment is now negatively associated with infant mortality though its explanatory ability remains quite small. It does suggest that for some infants, high tech treatments may be beneficial in lowering mortality rates. Finally, the availability of physicians increases in magnitude, from a -0.02 to -0.08 on infant mortality and from -0.04 to -0.06 for general mortality rates. This continues to suggest that having access to routine medical care is perhaps more important than having high tech treatment options or expensive diagnostic capabilities.

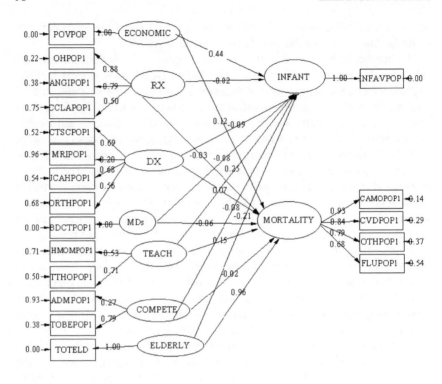

Chi-Square=543.50, df=121, P-value=0.00000, RMSEA=0.066

Fig. 5. Standardized Coefficients for full Structural Equation Model.

The final model demonstrates the interrelationships that exist between the constructs and the outcome variables. Infant mortality is correlated to general mortality and explains 29% of the variance in general mortality rates. Factors that influence general mortality also continue to exert an influence upon infant mortality. This final model clearly demonstrates that diagnostic services and treatment options have little ability to explain the variance in infant mortality or general mortality. While hospitals may be responding to pressures to decrease operating costs and increase technology, the influence on the well-being of the community is small. Physicians within the area continue to exert a positive influence by being negatively associated with mortality. Teaching hospitals perhaps because of their teaching mission or the type of patient they receive, are positively associated with infant mortality and general mortality.

CONCLUSIONS

I began this study with the desire to see the interaction between environment, healthcare services and mortality rates. While previous studies have focused on the hospital, their structure and characteristics, this study focused on the community as the unit of analysis, the community being operationally defined as healthcare service areas. The number of hospital beds, the number of CT scanners and the number of hospitals that provided high tech intensive services such as open-heart surgery demonstrate that hospitals are actively attempting to provide advance technological services. However, these same services and technologies contributed little to explain mortality rates within a healthcare service area. That finding suggests that the hospital industry responded to certain changes within the environment, such as cost containment, and, in order to survive identified the services and technologies that would generate financial stability. These changes in services and technology were developed or obtained because the insurance companies and/or Medicare would pay for these services and perhaps not necessarily to better serve the needs of the community or client. The sheer number of some technologies such as the number of MRI x-ray machines per population suggests that hospitals are obtaining technologies in response to something other than consumer need or demand. The findings of the structural models strongly suggests that factors such as poverty level, the number of elderly and the number of physicians within a service area have a greater effect on the outcome variables than variables associated with the hospital's technological services such as diagnostic ability and treatment options. These hospital level variables had little impact on the outcome variables and therefore upon the general well-being of the community. Availability of physicians and the economic resources to take advantage of routine or preventative medical care appears to have a stronger inverse relationship to infant and general mortality rates than high tech treatments and diagnostic services. This suggests that socioeconomic status, preventative prenatal care and routine medical care are more important than high tech treatments that a hospital can provide. When attempting to resolve the issue of rising health care costs, it becomes important to look at the entire system or ecosystem and not simply look at the parts within the system. This suggests that there are perhaps more salient issues for hospitals and healthcare services to address or that a redirection of services may be more important than the purchasing of the latest x-ray machine.

The model that looks at overall mortality issues suggests that there might be other variables that have more of an impact than the ones that I have developed. Communities that have a large number of the elderly will naturally have a higher level of mortality, but healthcare services, in theory, should at least promote some decrease in mortality rates. Determining which mix of services has the greatest impact on mortality rates should be studied. Further research should look at which preventative treatments, what routine diagnostic services or what aspects of a healthy lifestyle have the greatest impact upon mortality rates within a HCSA. Because the elderly population had such a strong relationship to mortality, perhaps looking at different age groups might provide some insight into which healthcare service variables are more important in decreasing mortality within the younger age groups. Other community attributes not specifically studied could be intervening in these findings and therefore suggest the need for further research.

NOTES

1. The ARF does not include 13 areas that were included in the original study. Whether these areas have been grouped into connecting areas is not clear. Each Hawaiian Island is provided a code and Honolulu has its own HCSA code.

2. Initially the death rate due to liver disease was also part of this measure but it had a large amount of error and a low loading score. The reason that liver disease had such a large amount of error and a low factor loading is probably due to the fact that direct death from liver disease is infrequent and that liver failure in many cases occurs due to other disease processes such as heart disease or cancer.

3. The covariance matrix was utilized to estimate all models presented. However to improve the interpretability of the models, the standardized parameter estimates are presented.

REFERENCES

Center for Disease Control (1995). Poverty and infant mortality: 1988. *Morbidity and Mortality Weekly Report, 44*(49), 922–925.

Fennell, M. L. (1980). The effects of environmental characteristics on the structure of hospital clusters. *Administrative Science Quarterly, 25*, 484–510.

Hannan, M. T., & Freeman, J. (1977). The population ecology of organizations. *American Journal of Sociology, 82*, 929–964.

Hartz, A. J., Krakauer, H., Kuhn, E. M., Young, M., Jacobsen, S. J., Gay, G., Muenz, L., Katzoff, M., Bailey, R. C., & Rimm, A. A. (1989). Hospital characteristics and mortality rates. *New England Journal of Medicine, 321*(25), 1720–1725.

Horbar, J. D., Badger, G. J., Lewit, E. M., Rogowski, J., & Shiono, P. H. (1997). Hospital and patient characteristics associated with variation in 28 day mortality rates for very low birth weight infants. *Pediatrics, 99*, 149–156.

Kuhn, E. M., Hartz, A. J., Krakauer, H., Bailey, R. C., & Rimm, A. A. (1994). The relationship of hospital ownership and teaching status to 30 and 180 day adjusted mortality rates. *Medical Care, 32*(11), 1098–1108.

Makuc, D. M., Haglund, B., Ingram, D. D., Kleinman, J. C., & Feldman, J. J. (1991). Health service areas for the United States. *Vital and Health Statistics, 11*, 1–102.

National Center for Health Statistics (1991). National Maternal and Infant Health Survey. U.S. Department of Health and Human Services Public Health Service.

Perrow, C. (1986). *Complex organizations: A critical essay* (3rd ed.). New York: Random House.

Pollack, M. M., Cuerdon, T. T., Patel, K. M., Ruttimann, U. E. (1994). Impact of quality-of-care factors on pediatric intensive care unit mortality. *Journal of the American Medical Association, 272*(12), 941–947.

Shortell, S. M., & Hughes, E. F. X. (1988). The effects of regulation, competition and ownership on mortality rates among hospital inpatients. *The New England Journal of Medicine, 318*(17), 1100–1107.

Thorpe, K. E. (1997). The health system in transition: Care, cost and coverage. *Journal of Health Politics, Policy and Law, 22*(2), 339–361.

THE HISTORY OF NURSING HOME
BED SUPPLY IN CHICAGO: THE
EFFECT OF FEDERAL POLICY AND
URBAN SETTLEMENT ON
UTILIZATION

Susan C. Reed

ABSTRACT

In the past decade, national nursing home utilization rates for African Americans have risen above those of White elders suggesting improved access to care. This study examines the effect of Medicaid upon the supply of long-term care facilities in Chicago communities by tracing the construction and placement of homes at three points in the development of federal long-term care policy compared to the settlement and segregation of the city's neighborhoods by ethnoracial groups. Spatial analysis of nursing home distribution shows why facilities built after 1965 are more likely to serve African Americans. The policy implications of changing long-term care utilization patterns are discussed.

Changing Consumers and Changing Technology in Health Care and Health Care Delivery,
Volume 19, pages 103–130.
Copyright © 2001 by Elsevier Science Ltd.
ISBN: 0-7623-0808-7

INTRODUCTION

Historically, the typical nursing home resident has been an older white woman with utilization limited, for elders of color, by patterns of discrimination and supply. However, by the end of the 20th century, national statistics show African Americans leading white elderly in the consumption of nursing home beds. Access to care for African Americans has improved as the result of important federal legislation such as the Civil Rights Act of 1964, which encouraged the integration of hospitals; and the passage of Medicaid in 1965, which provided funding for both acute and long-term care of low-income persons (Smith, 1999). However, shortages of Medicaid beds remain in some geographic areas, indicating that state policies and local health care markets complicate the implementation of federal policy to improve access (DuNah et al., 1995; Ginsberg et al., 1993). The passage of Title 19 of the Social Security Act coincided, in the mid-1960s, with dramatic changes in the composition of urban neighborhoods as suburbs attracted middle class families and the poor were concentrated in low income communities. The push and pull of federal and local trends have worked together to increase the supply of nursing home beds to some African American communities while leaving an inadequate supply in others.

This study of Chicago's long-term care history over the course of the past century explores the effect of Medicaid upon the construction and location of nursing homes. By analyzing the increase or decrease in bed supply subsequent to certain key federal policy events, and considering changing locations of nursing homes relative to racial residence, the effect of both federal policy and urban settlement patterns are considered. The objective is to identify factors that contributed to success in improving access to nursing home care despite persistent racial segregation in urban communities (Anderson & Massey, 2001) and nursing facilities themselves (Wallace, 1990; Falcone & Broyles, 1994; Smith, 1999).

Studies of Medicaid's effectiveness show significant increases in the utilization of health care services, including nursing homes, for Blacks and Latinos since the 1960s. In 1963, 10/1000 Black elderly resided in a nursing home compared to 27/1000 White elderly (Smith, 1999). According to the 1997 National Nursing Home Survey, 54/1000 Black persons over 65 are nursing home residents, compared to 44/1000 White persons of the same age (National Center for Health Statistics, 2000). However, Latinos remain less likely than either group to reside in a nursing home (Clark, 1997). Twenty-six percent of Whites over 85 lived in nursing homes in 1990, compared to 17% of African

Americans and 10% of Latinos of the same age (Damron-Rodriguez et al., 1994).

Some scholars argue that cultural differences in the expectation of care for the elderly are responsible for lower supply and utilization in the Latino community, while others consider structural barriers to be the cause of the disparity (Angel et al., 1996; Wallace et al., 1998). Latino elders are less likely than either African American or White elders to live alone (U.S. Bureau of the Census, 1996) but are also less likely to participate in Medicaid. Eighteen pecent of Latinos receive Medicaid while 25% of African Americans participate in the federal program (Angel & Angel, 1997). As Medicaid is the only program that covers long-term nursing home care,[1] communities with few private pay patients and low Medicaid participation may be less likely to attract nursing home providers (Angel & Angel, 1997). In Chicago, Latino neighborhoods have a lower supply of nursing home beds per elderly (Reed & Andes, 2001).

Studies of long-term care utilization rates by race and ethnicity show variation by metropolitan area as well as ethnicity. Wallace (1993) found that while whites dominated nursing home residence in his study of 40 metropolitan areas, in one quarter of U.S. cities African Americans had higher utilization rates than whites. Other studies have found higher rates of utilization among African American elders than would be predicted by their proportion in the City's elderly population (Douglas et al., 1988). In Chicago, 36% of nursing home residents are African American compared to only 30% of elders (Reed et al., 2001).[2]

In Wallace's study (1993), African American utilization of nursing home beds was related to higher levels of racial segregation among the city's nursing facilities. This finding raised the question, do characteristics of a city's residential housing patterns affect the utilization of long-term care beds? Racial segregation would be expected to block African American utilization if homes are most plentiful in white communities. However, there is some evidence that segregation might actually increase the supply of beds in some urban markets by reducing competition from private pay consumers (Reed et al., 2001).

Nursing home supply theory appears to support the relationship between the segregation of Medicaid patients and the supply of nursing home beds. Scanlon (1980) and Nyman (1993) found that Medicaid beds are constrained by the availability of private pay patients. Nursing home operators will admit private-paying patients first and fill the remaining beds with residents whose care is reimbursed by Medicaid. Ettner (1993) found that in counties where private pay demand is high and bed supply is low, Medicaid patients wait for admission. The reverse of this theory may be at work in Chicago's poor communities where bed supply is high and private pay demand virtually eliminated by economic segregation. Where Medicaid eligible persons are concentrated in large sections of the City, nursing home operators specialize in Medicaid reimbursed care increasing the supply of long-term care beds.

Wolinsky (1990) identified three historical periods that had a significant effect upon the utilization of long-term care by older Americans: (1) In 1935, the passage of Social Security launched a generation of long-term care consumers by providing a minimum income for seniors, allowing some who would otherwise have gone to the poorhouse to pay the entrance fee to a home for the aged. While such homes were racially segregated, a limited number of such facilities were available to elders of color; (2) In 1965, the passage of Medicare and Medicaid allowed elders who had been impoverished by growing medical bills and largely considered undesirable patients, to obtain care from physicians, hospitals and a growing number of nursing homes that were adopting a medical model. As recipients of federal funds, hospitals were forced to integrate but the desegregation of nursing homes was never enforced; and (3) The Social Security Amendments of 1983, established a new reimbursement system for Medicare, commonly referred to as DRGs. The result was a reduction in hospital length of stay and an increase in the need for long-term care for elders of all races, some of which was met by Medicare's home health coverage and hospitals' conversion of beds to skilled nursing units.

This paper will trace the supply of nursing home beds in Chicago at each of these three periods showing differences in their location that relate to racial patterns of utilization. As whites moved to the suburbs in the 1960s, nursing homes that had been established for a particular ethnic group were left behind to serve an increasingly African American community. With the help of Medicaid funding, this transition may have increased the supply of beds to a community that might otherwise have remained unserved. In addition, the desegregation of hospitals caused the closing of African American hospitals and, in other cities, their conversion to nursing facilities for communities of color (Smith, 1999). We will analyze Chicago's history to determine whether the congruence of racial residential change and the passage of Medicaid and Medicare resulted in these and similar events that contributed to the increased utilization of long-term care services by persons of color throughout the past century.

Spatial analysis can illuminate national utilization trends to reinforce commitment to effective federal policy as well as identify remaining barriers to access for those groups and communities that remain underserved. As alternatives to nursing home care are developed in cities and states all over the country, care must be taken to ensure that the distribution of these new services is equitable. The consideration of residential housing patterns is required so that nursing homes do not become the only long-term care service available to elders of color.

METHODS

Measurement

Analyses of two directories, *The Social Service Directory* (published first by the Welfare Council of Metropolitan Chicago and then the United Way) and the *City Directory of Chicago*, provided data on existing facilities at each decade of the 20th Century. The Welfare Council of Metropolitan Chicago published directories of not-for-profit "Homes for the Aged" and Hospitals going back to, 1915, with information about the year in which the Home was established. "Nursing Homes" (mostly proprietary), were only listed in the City Directory of Chicago allowing us to distinguish between the two types of ownership. These data were analyzed to determine what year existing homes were established, how property changed hands over time, whether hospitals became nursing homes and whether not-for-profits became for profits.

After analyzing changes in Chicago's nursing home and hospital supply across the century, 101 skilled and intermediate care facilities located within the central city are divided into two groups – those built before 1965 and those constructed after the passage of Medicaid. Data for this analysis was drawn from the *1990 U.S. Census of Population* and Illinois' 1994 *Long Term Care Facility Survey*. Differences between the two groups are examined on the following measures: *percent African American residents of the facility, percent Latino residents of the facility, percent Medicaid reimbursed care, percent African American residents of the census tract, percent facilities located in an area of concentrated poverty,*[3] *percent facilities for-profit.*

Hypothesis

It is expected that patterns of racial resettlement throughout Chicago's history will affect the supply of nursing homes to different ethnoracial neighborhoods. However, Chicago's experience may vary from that of other cities where, for instance, Black hospitals were converted to nursing homes in African American communities. It is predicted that the majority of homes built after 1965 will be for-profit facilities with high percentages of Medicaid residents who are African American. It is further anticipated that these facilities will be located in areas of concentrated poverty that are predominately African American.

Chicago as Study Locale

Some aspects of Chicago's development are unique among American cities and so it's important to consider whether the following history can be generalized to that of other urban areas.

Although Chicago has a distinctive history in the depth of racial antagonism as well as the strength of the Democratic machine, it is a typical American city in many aspects that relate to the distribution of long-term care making it an appropriate site for this study. Chicago leads most other cities in the level of racial and economic segregation but measures such as population living in extreme poverty declined between 1980 and 1990 while increasing in many other U.S. cities (Kasarda, 1993; O'Connor et al., 2001). Likewise, Chicago had a particularly powerful Democratic machine that influenced the growth of not-for-profit homes for the aged early in the century as a means of party building and encouraged (but sometimes discouraged) the development of proprietary long-term care establishments. However, a similar interaction between the growth of political parties and the evolution of social services has been well documented in most areas of the country (Skocpol, 1992) and will be discussed below.

In general, the city's age may have an effect on the development of nursing home supply relative to cities that developed both earlier (eastern cities) and later (western cities). Chicago's concurrent population growth and industrialization may have contributed to the strength of its neighborhoods and their ethnic solidarity. Its youthful, entrepreneurial local economy may have encouraged the growth of proprietary long-term care within these communities while the sector in older cities remained not-for-profit (Ginsberg et al., 1993).

State policy also influences the rate and timing at which a city's long-term care infrastructure is built. Illinois has remained fairly typical of other states throughout the century in both its funding of nursing homes and its efforts to regulate the industry. Although Illinois established a Certificate of Need program in the 1970s to control overbedding, the state has more nursing homes than most other states. This would suggest an environment conducive to nursing home interests. Indeed, owners organized a powerful lobby for reimbursement rates that would sustain profitability (Chicago Tribune, 1979, 1980). However, Medicaid eligibility and expenditures in Illinois have continually hovered about the national average (Long & Liska, 1998). Although Illinois' supply of beds per person over 65 has remained higher than the national average for several decades, the rate of growth between 1978–1993 is lower suggesting that the development of Illinois long-term care infrastructure occurred earlier than some other states (DuNah et al., 1995). Much of this early growth occurred in Chicago rather than suburban or more rural areas of the state.

BEFORE 1935: CHARITABLE CARE AND SOCIAL CONTROL

By the time of Chicago's rapid growth in 1840, the almshouse had emerged on the East Coast as the main form of charity to the poor. Chicago adopted the model but termed the institution the "poorhouse" (Gintzig et al., 1969). The Cook County Board of Commissioners built Chicago's almshouse in the center of the city in the area that is now called the Loop with additional poorfarms on the outskirts of town. In 1848, the commissioners tried to replace financial support in the home entirely with this institutional form of charity but by 1859, this decision had been overturned (Rothman, 1971).

The composition of these facilities reflected the relative social standing of different ethnic groups. By 1890, 79% of Chicago's 1 million people were first or second generation immigrants including Irish, Germans, Poles, Swedes, Czechs, Dutch, Danes, Norwegians, Croatians, Slovaks, Lithuanians and Greeks. The black population of Chicago tripled in 1850 after the passage of the Fugitive Slave Law that made Chicago an important stop on the Underground Railroad. Latinos, primarily Mexicans and Puerto Ricans, did not come to Chicago in large numbers until WWII (Cutler, 1982). Of the 610 poorhouse residents in Chicago, 92% were foreign born. Fifty-three percent of Chicago's almshouse residents were Irish at a time when they comprised only 13% of the city's population (Rothman, 1971).

Conditions in Chicago's poorhouses were similar to those in New York and Massachusetts – deplorable. Rothman (1971) quotes a grand jury report of 1853 finding the institution severely overcrowded with beds stuffed into every corner and poorly ventilated. The sick and the well roomed together resulting in unhealthy and unpleasant living conditions. However, there was little sentiment in favor of improving the environment of the disabled poor. Fear of the poorhouse was thought to prevent indolence and rebellion among the working poor in the face of equally undesirable working conditions.

Most residents of the poorhouse were not able bodied although the almshouse was meant originally to provide work for the idle. Surveys in the 1870s found that they were inhabited mainly by immigrants who were mentally ill (although mental asylums were sprouting in every state) or disabled elders and children (although orphanages were the preferred form of care for destitute children) (Katz, 1986). By 1920, 70% of almshouse residents were over 55 years of age. This proportion reflected a growing need for care of the increasingly destitute elderly (Vladek, 1980).

Since colonial times, individuals, often unemployed nurses, had taken a few elderly or disabled people into their homes for income. But at the turn of the century there was not sufficient capacity to care for the elderly whose proportion in the population was growing about 1% a decade (Vladek, 1980). To provide an alternative to the indignity of the poorhouse, ethnically identified religious and voluntary associations began to build homes for the aged. Nineteen such homes were built within the City of Chicago before 1915. Most served the elderly of a specific ethnic group and religion. For instance, German Jews who settled the neighborhood of Hyde Park in the 1870s built Michael Reese Hospital in 1867 and the Home for Aged Jews in 1893. Russian Jews who came to Chicago the following decade settled in the near West Side and built the Otrthodox Jewish Home for the Aged in 1899. Other homes built that same decade were the Augustana Home for the Aged sponsored by the Swedish Evangelical Church, the Bohemian Old People's Home and Orphan Asylum ("Home and shelter for old people of Bohemian nationality") and the Danish Old People's Home.

Home-like alternatives to the almshouse were not as available for the disabled of other age groups. Only a few homes were built toward the end of the 19th century for "destitute crippled children". The Central Free Dispensary at Rush Medical College was established by the State in 1870 "to house and care for the feeble minded children of Illinois." The treatment of the mentally ill which had become the purview of the State rather than County government was remanded to large state mental hospitals of which there were ten in Illinois by the turn of the century.

Religious groups were sponsoring the building of private hospitals for the first time as well. Until this time, hospitals, such as Cook County Hospital in Chicago and Bellevue in New York were strictly charities that were affiliated with almshouses for the medical care of the poor. Those with means preferred to receive medical care at home where the risk of infection was much less (Katz, 1986). However, in Chicago the German American Hospital, Swedish Covenant Hospital and Michael Reese Hospital were established to serve particular ethnically identified religious groups although many advertised (as early as 1915) that they were non-sectarian and served the sick without regard for race, creed or color.

The African American community also established health care institutions during this period. Many Black leaders discouraged separate social institutions as segregationist. However, as white hospitals, homes for the aged and even YMCAs refused admission to African Americans, the black elite began to organize a hospital and home for the aged in the community. Provident Hospital was founded in 1891 by prominent physician, Daniel Hale Williams, who envisioned an integrated medical facility for physicians and residents of both races (Spears, 1961). An African American women's club founded the Home for Aged and Infirm Colored People on Garfield Boulevard, with wide support from the community providing supplies and funding from prominent Black businessmen and women of means (Knupfer, 1996). By 1930, another home for the aged, called the African ME Deaconess and Stewardess Home would be established in the community.

The construction of these ethnically identified homes for the aged resulted from the friendly relationship between ethnic group leaders and Chicago's Democratic Machine, which, together with the state legislature was happy to distribute land and other resources to those with a broad voting constituency. In this, Chicago was not unique. Throughout the country in this era of party building local governments created social policy by distributing government funds to the worthy voting poor (Skocpol, 1992). By supporting Catholic homes for the aged, Irish politicians began to build a base among those seeking to save their loved ones from the fate of the poorhouse.

The racial segregation of homes for the aged had been firmly established and reflected growing opposition to the expansion of the African American community. While whites and blacks lived together on many blocks on the south side until 1910 when African Americans comprised less than 2% of the city's population, tensions escalated in the subsequent decade as the proportion of African Americans doubled (Drake & Cayton, 1945) culminating in the Riot of 1919 in which 38 people were killed (Spear, 1967). By 1930, separation of the races was so well established, that when a University of Chicago graduate student asked representatives of Homes for the Aged if African Americans were admitted, most responded that the issue never came up. When a White person sought admission to the Home for Aged and Infirm Colored Persons, the Chicago Chamber of Commerce opposed the integration arguing that it would set a bad precedent (Glick, 1930). Throughout the century many homes for the aged would expand both services and size, however, homes in the African American community were not the beneficiaries of large bequests so remained separate and unequal.

Prior to the passage of Social Security in 1935, life expectancy was less than 60 years[4] and more than one-third of the poor were over 65 years of age (Gornick et al., 1996). Then, as now, families cared for the vast majority of disabled older persons. Homes for the aged offered an alternative to the poorhouse for older adults who could raise the admission fee, which averaged $500 in 1930 along with the transfer of all assets (Glick, 1930). Before 1935, disabled older adults were not viewed as consumers of long-term care. Rather they were the beneficiaries of charitable care as residents of homes for the aged and recipients of social control as inmates of the poorhouse (Cohen & Scull, 1983).

SOCIAL SECURITY: THE ELDERLY BECOME CONSUMERS OF LONG-TERM CARE

With the Depression of the 1930s, came destitute circumstances for many of the 7 million elderly living in the United States. In response, the Social Security Act was passed in 1935 to provide a minimum of financial support for older citizens who had fallen on bad times.[5] As the law forbade money to be given to anyone living in a public institution, those elderly living in almshouses (and twice that number living in state supported mental hospitals) were ineligible (Vladek, 1980) while those living in homes for the aged were qualified for government assistance. As Congress intended, support for the poorhouse began to fade and the development of more home-like alternatives increased.

While proprietary boarding homes had always existed as residential options for the elderly, those that marketed directly to this population proliferated after the Social Security Act provided an income to those otherwise without means. By 1946 more than 40 new enterprises of this type were operating in Chicago neighborhoods. Nurses and other health professionals opened the doors of large houses to paying customers: Dr. Gomberg Convalescent Home on Drexel Blvd., Helen Gould Rest Home on South Wood, Elsa S. Long Convalescent Home on Sheridan Road. Although the make-up of their communities undoubtedly influenced the ethnic composition of these homes most had no clearly stated ethnic identification. The exceptions were the German Old People's Home, the Jewish Old People's Home and the Open Door Gospel Home which were proprietary with ethnically identified names rather than not-for-profits built by an established religious organization. These small board and care homes co-existed with larger ethnically based facilities that had been open since the turn of the century.

Blacks resided in separate boarding homes from whites and utilized county facilities but these facilities were overcrowded and the need for long term care among Blacks was largely unmet (Smith, 1990). To increase the supply of beds to the poor aged, local welfare commissioners contracted with private boarding homes. In Chicago, only the Montgomery Convalescent Home on South Prairie advertised that it welcomed "colored aged". Members of the African American community working hard to keep open the Home for Aged and Infirm Colored People, purchased a property in Chicago's Bronzeville community (officially called Grand Boulevard) to expand the number of elders who could be admitted.

Gradually, real estate speculators entered the nursing home business and Congress responded with regulation and limited funding. These entrepreneurs acquired land and saw conversion to a nursing home as relatively effortless and profitable use. From the beginning, conditions in these homes were deplored and licensing was called for (Zinn, 1999). In 1950, Congress added three nursing home provisions to the Social Security Act: the prohibition against payments to those in institutions was removed; states could provide payment to vendors directly and receive federal match; and licensing was required (Institute of Medicine, 1986). Although the construction of thousands of hospitals was financed in 1946 with the Hill Burton Act, Congress was reluctant to fund the private nursing home industry. Despite lobbying efforts by the newly formed American Nursing Home Association, Congress decided to limit such funding to not-for-profit or public facilities in the 1950s. Not-for-profit nursing homes were required, under this legislation, to operate in conjunction with a hospital while for-profits, that were not eligible for funding, were not (Vladek, 1980).

Despite federal encouragement of not-for-profit and public nursing homes, the growth of for-profit homes far outpaced that of not-for-profits in Chicago. Approximately 90 new for-profit homes were opened by 1955 (while 30 ceased operation between 1945–1955) and only three new not-for-profits were opened. Ethnically-identified and religious organizations were more concerned with the construction of hospitals (Vladek, 1980) eight of which were constructed in Chicago between 1946–1956. Since states, at this time, were enacting minimal regulation that was loosely enforced, private nursing homes continued to have the appearance of homes rather than health care facilities as they were still located in old mansions and boarding houses often bought by investors from the original owner-operator.

In 1945, the Institute of Medicine called a meeting in Chicago to discuss state supervision of Illinois' nursing homes. The result was the passage of the Nursing Home Licensing Act in that year which gave the responsibility to a reluctant Illinois Department of Public Health. In the next year, 95 licenses were issued and the Department had begun working with providers to improve fire safety and staffing. However, some operators fought regulation, leading the Director of IDPH to comment,

> Because of economic and political ramifications, the licensing of these homes, with an eye on conformity with minimum standards as clearly defined by law, is perhaps the most sensitive operation in the Department (Richardson, 1963, 125).

Nursing home operators were gaining political influence as evidenced by the passing of an amendment (in 1957) that called for the establishment of The Advisory Council on Nursing Homes and required that four of the eleven members be representatives of provider associations (Richardson, 1963).

When the burgeoning nursing home industry finally received federal support for new construction of homes, growth was experienced mainly in Chicago's suburbs where new construction of facilities was necessary, as old housing was less available (Hospital Planning Council for Metropolitan Chicago, 1968). The Hill Burton Act was amended (1954) to include funds for nursing homes and the American Nursing Home Association won loans from the Small Business Administration (1956) as well as FHA (1959), which authorized almost a billion dollars in nursing home loans across the country (Wittendorn, 1983; Institute of Medicine1986). In the City, there were 40 new facilities but roughly the same number of nursing homes closed their doors between 1955–1965 including several not-for-profits that built newer facilities on the north side where their respective ethnic groups were moving.

By the 1960s, the location of the City's nursing homes began to shift to the north side. The late 1950s and early 1960s saw the growth of the Black Metropolis along with "white flight" from the south and west side communities that European ethnic groups had built since first settling in Chicago. Between 1950 and 1980, nearly one fifth of Chicago's population, mostly white, left the central city (Squires et al., 1987). Many white families remained in the City but moved in greater numbers to the north side. While in previous decades, homes for the aged were constructed all over the City, the majority of those that closed between 1955–1965 were on the south and west sides, primarily African American and Latino neighborhoods, while most of the new facilities were opened on the north side of the City (Hospital Planning Council for Metropolitan Chicago, 1968).

To some extent during this decade, African American families could utilize the nursing homes and hospitals that had been built earlier in the century. Although some not-for-profits moved to the north side in this decade, others remained on the south and west sides. The Orthodox Jewish Home for the Aged remained in North Lawndale although the Jewish community had left; and the Home for Aged and Infirm Colored People was renamed the Jane Dent Home and continued to operate as a not-for-profit until the 1990s. In several cases, homes were opened on the site of hospitals that had closed. Communities in which Latino families were settling, however, experienced several closures without new construction of nursing homes (Hospital Planning Council for Metropolitan Chicago, 1968).

Between 1935 and 1965, the nursing home industry grew in number of facilities and influence over state and federal policy. Demand for nursing home care was fed by rising numbers of older adults as well as a measure of financial independence awarded by the establishment of a national pension system. Older adults and their families began to see themselves as consumers of long-term care in response to increasingly market relations and the decline of charitable care. However, the most desirable clientele were those with more than their social security check to spend. As the spatial distribution of wealth around the

City began to change, so did the supply of long-term care beds, both for-profit and not-for-profit. The fact that coverage of medical services was not included in the original Social Security legislation meant that escalating health care expenses were continuing to impoverish many elderly. The inequity of health care services and pressure from providers on Congress to insure the elderly poor would lead to a dramatic shift in the federal government's involvement in funding for long-term (and acute) care (Kronenfeld, 1997).

MEDICAID: LOW INCOME CONSUMERS ENTER THE MARKET

Between the growth of the health care industry, including nursing homes nationwide, and the rapid increase in the number of elderly disabled persons, the pressure for national health insurance was mounting. In 1960, the Kerr-Mills Act was enacted with several provisions that would affect the subsequent design of Medicaid's long-term care coverage. Kerr-Mills was passed to provide medical assistance for the indigent aged. States were permitted to cover skilled nursing services, loosely defined; to cover those who were impoverished by high medical expenses; and to receive a relatively high federal match for state payments. As a result, some nursing homes were receiving significant federal funds before Medicare and Medicaid were passed. In fact, vendor payments nationally increased more than $400 million between 1960–1965 (Vladek, 1980).

After the defeat of expanded health insurance for the elderly during the Kennedy administration, the Johnson administration continued to fight for Medicare, an amendment to the Social Security Act that would fund physician and hospital care for all elderly but only limited skilled care in nursing homes. Medicaid was designed as an expansion of the Kerr Mills Act to the non-elderly poor so nursing home care was included without the restrictions on the duration or type of care included in Medicare. Most commentators argue that in their haste to pass health insurance for the poor as well as the elderly, reformers ignored the inconsistencies between Medicare's and Medicaid's approaches to long term care (Institute of Medicine, 1986).

After passage of Titles 18 and 19 of the Social Security Act, states scrambled to update regulations and increase staffing in order to ensure compliance with new standards set by Medicare. In Illinois, engineers were overwhelmed with the task of supervising new construction so a job classification called "geriatric home advisors" was created (Wittenborn, 1983). In the meantime, claims were denied for facilities that had been certified but then did not meet the required list of services. Such facilities dropped Medicare and families were stuck with thousands of dollars in bills (Vladek, 1980). Older facilities in Chicago faced a high rate of under compliance.

The Illinois Department of Public Health found any construction that was over 40 years old "non-conforming". In the City, only 22% of the 108 facilities operating in 1966 were considered adequate by national standards while in the suburbs, 50% (of 144) were in compliance. Seventy-eight percent of the City homes that did not comply had fewer than 50 beds suggesting that many had formerly been residences converted to nursing home care. Black south side and near west side communities were more likely to have substandard facilities than other communities of the City, which may again be attributed in part to the age of the structures. Only 11% of beds in these neighborhoods met the standard compared to 40–60% in the other communities of the City (Hospital Planning Council for Metropolitan Chicago, 1968).

In order to avoid forcing all nursing homes to conform to high medical standards, the Intermediate Care Facility (ICF) was created (under the Miller amendment). Standards were not set for these ICF's until 1971 and some states merely reclassified their substandard facilities. In 1971, standards were set that made them similar to Skilled Nursing Facilities, which remained more highly regulated. The Social Security Amendments of 1972 allowed states to cover Intermediate Care Facilities under Medicaid (Vladek, 1980).

While the passage of Medicaid and Medicare stimulated substantial growth in nursing homes in both the City and the suburbs, homes that had previously operated in Chicago were forced to either build new structures or go out of business. Smaller boarding homes died out because of the expense of complying with stricter regulations such as sprinklers that were required in Illinois in 1972 (Wittenborn, 1983). In some cases, joint partnerships were formed as those able to comply with regulations bought out others (Vladek, 1980). While 32 new nursing homes were opened in the City between 1965–1975, 60 facilities that had been operating in 1965 were out of business by 1975. This did not result in a decline in nursing home beds, however, as the new facilities were larger, replacing those that had occupied the City's older mansions.

Health care institutions receiving Medicare and Medicaid funds were required to comply with the Civil Rights Act of 1964 that forbade discrimination "under any program or activity receiving federal assistance". Medicare and Medicaid, as Smith (1999) shows, would make health care a less financially risky venture and the providers only needed to integrate their facilities to receive these federal funds. However, since Medicare would not cover most nursing home care and since Medicaid payments were from the beginning below cost, nursing homes had less motivation to desegregate than did hospitals. At the same time, the federal government did not enforce compliance with the Civil Rights Act among nursing

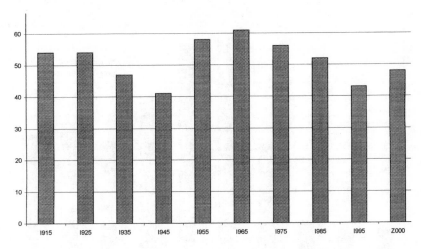

Fig. 1. Number of Hospitals in Chicago 1915–2000.

Sources:
Chicago Department of Public Welfare. 1915. *Social Service Directory*.
Welfare Council of Metropolitian Chicago. 1926. *Social Service Directory, Chicago*
Welfare Council of Metropolitian Chicago. 1936. *Social Service Directory, Chicago*
Welfare Council of Metropolitian Chicago. 1946. *Social Service Directory, Chicago*
Welfare Council of Metropolitian Chicago. 1956. *Social Service Directory, Chicago*
Welfare Council of Metropolitian Chicago. 1963. *Social Service Directory, Chicago*
United Way/Crusade of Mercy. 1977/1978. *Social Service Directory, Metropolitian Chicago*
Illinois Department of Public Health. 1994. *Illinois Long–Term Care Facility Survey*
Chicago City Directory. 1915, 1926
Chicago Telephone Directory. 1936, 1945/1946, 1955, 1965, 1975, 1977, 1985, 1995, 2001.

homes. Smith (1999) reports that the Johnson administration was less inclined to tell nursing home residents how to live. Such a policy came uncomfortably close to the desegregation of housing which had never been broached on either the federal or local level. Wallace (1990) shows that, in St. Louis, segregation levels of nursing homes rose after 1961 almost achieving the levels of residential areas by 1988.

The desegregation of the nation's hospitals put financial pressure on the nation's 124 Black owned hospitals as Black physicians and patients were recruited away to integrating white hospitals (Smith, 1999). Some such Black institutions closed after holding on for a decade or two; others, including a Black owned hospital in Detroit, were converted to a nursing home (Douglas et al., 1988). In Chicago, Provident Hospital continued to operate through the sixties but closed for a period of time before reopening in the 1990s as part of the Cook County Hospital network.

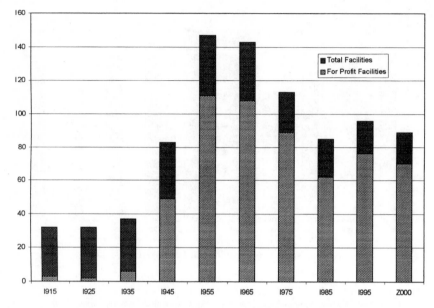

Fig. 2. Number of Nursing Homes in Chicago: 1915–2000.

Note: By 1995, 11 skilled nursing units were either attached to or located within hospitals. These facillities are not included in this analysis while shelter care homes are included because of their comparability to homes that existed earlier in the century.

Sources:
Chicago Department of Public Welfare. 1915. *Social Service Directory*.
Welfare Council of Metropolitian Chicago. 1926. *Social Service Directory*.
Welfare Council of Metropolitian Chicago. 1936. *Social Service Directory*.
Welfare Council of Metropolitian Chicago. 1946. *Social Service Directory*.
Welfare Council of Metropolitian Chicago. 1956. *Social Service Directory*.
Welfare Council of Metropolitian Chicago. 1963. *Social Service Directory*.
United Way of Metropolitian Chicago. 1977/1977. *Social Service Directory*
United Way/Crusade of Mercy. 1986. *Care Service Directory of Metropolitian Chicago*
Illinois Department of Public Health. 1994. Illinois Long-Term Care Facility Survey.
Chicago City Directory. 1915, 1926
Chicago Telephone Directory. 1936, 1945/1946, 1955, 1965, 1975, 1977, 1985, 1995, 2001

Provident Hospital's experience is common to many community hospitals in Chicago, especially those located in increasingly poor communities that closed their doors as the result of mounting financial pressure. Whiteis' (1992) study of hospitals nationwide found that the percentage of black residents in the community predicted closure. Eighteen Chicago hospitals[6] closed their doors between 1965 and 1995, leaving the poorest communities with few alternatives to the public hospital system. To this day, very few hospital-based skilled nursing units are located in Chicago's African American communities (Salmon, 1993).

Inequity in the distribution of health care facilities, prompted legislation to require states and local communities to plan investment more systematically. The Health Planning and Development Act of 1974 reflected a community health planning perspective that had been suggested by Hill Burton but without any specific formula for local areas to follow (Budrys, 1986). The new law, on the other hand, established 205 Health System Agencies with the responsibility of developing a plan to improve access in the community by making the construction of facilities located in medically underserved areas a priority and promoting investment to provide a continuum of care. To this end, each state was required to establish a certificate of need (CON) process with review boards on which consumer representatives were the majority.[7] Unfortunately, this legislation did not forestall hospital disinvestment in communities of color. H.S.A. Boards could only refuse permission to develop, not initiate development or support facilities in low-income communities with additional resources (Budrys, 1986).

Nursing home construction in the south and west sides, on the other hand, did increase during this time. Because Medicaid legislation included provisions to cover capital costs, long-term care facilities were a profitable real estate investment, especially since each time the facility changed hands capital costs were recalculated at a higher value[8] (Goldberg, 1995). Furthermore, the concentration of poverty that was coming to characterize African American communities meant that a large proportion of the nursing home market in such communities would now be eligible for long-term care insurance under Medicaid allowing providers to "specialize" in Medicaid care.[9]

Most of Chicago's nursing homes located in areas of concentrated poverty were built after 1965, when Medicaid was launched as shown in Table 1. In fact, nursing homes built since 1965 are significantly more likely to be for-profit enterprises, located in African American communities and to serve African American residents. Medicaid reimburses 79% of the care that these facilities provide. This analysis indicates that Medicaid funding has stimulated the supply of nursing home beds for long-term care consumers that might otherwise have remained outside of the nursing home market by virtue of their low income.

Concern remains, however, about the quality of the care provided in large Medicaid nursing homes. With Medicaid reimbursement rates below the cost of care in every state, providers operate with fewer staff and amenities for larger numbers of residents per facility in order to realize an economy of scale and higher profits. Studies of facilities that are reimbursed primarily by Medicaid have found lower expenditures (Nyman, 1988) lower levels of RN staffing

Table 1.   Characteristics of Chicago's Nursing Homes, Constructed
Before and After 1965. $N = 101^i$

	Before 1965 $n = 49$	1965 or later $n = 52$
Mean %		
Black in facilities' census tracts**	24	42
Black residents**	22	40
Latino residents	4	3
Medicaid reimbursed**	57	79
Percent of homes (No.)		
For-Profit*	42 (30)	58 (41)
Located in area of concentrated poverty*	22 (4)	78 (14)

Source: 1990 U.S. Census of Population; 1994 Illinois Long-Term Care Facility Survey

* Differences between groups significant, $p < 0.05$
** Differences between groups significant, $p < 0.01$

[i] Included in this analysis are skilled and intermediate care beds, including 11 facilities that are attached to or located within hospitals. Skilled beds provide 24 hour nursing care certified by both Medicare and Medicaid. Intermediate care facilities provide basic nursing care certified by Medicaid only. Not included in this analysis are 27 homes that are sheltered care, facilities for those 22 and under and facilities for persons with developmental disabilities. One facility is missing data on Medicaid and race/ethnicity.

(Nyman, 1988; Zinn, 1994) and indicators of worse patient outcomes by such measures as percentage of residents not toileted (Zinn, 1994). In Illinois, both the Department of Public Health and the Department of Public Aid monitor the care provided in Medicaid homes but are reluctant to close those facilities that provide much needed care to poor communities (Wittenborn, 1983).

In Chicago, Latino persons are not as likely as African Americans to utilize Medicaid beds. Nor have Latino communities in the City benefited from the increase in bed supply stimulated by the passage of Medicaid. On the other hand, Latino elders who do utilize nursing home care are not as restricted by racial segregation as are White and African American persons. In a previous analysis (Reed & Andes, 2001), the author found that Whites and Blacks were unlikely to occupy nursing homes in each other's communities but Latino elders utilized nursing homes in either community.

The passage of Medicaid in 1965 coincided with significant demographic changes in urban neighborhoods that resulted in the construction of new nursing home facilities in Chicago's African American communities. The closure of hospitals in those same communities and reduction in the supply of physicians

(Fossett & Peterson, 1989) made nursing home care one of the few medical alternatives for African American elders and their caregivers. The high cost of nursing home care continued to restrict access for many individuals and communities of the City who could not meet Medicaid's strict income criteria for eligibility. For some older adults, however, the passage of Medicaid made the utilization of long term care services a reasonable, if undesirable, residential alternative for later life. As this expectation spread, so did consumer activism to improve the quality of nursing home care. The Health Planning and Development Act had laid the groundwork for consumer involvement to improve access to care; the next step would be to include consumers in decisions that would improve the quality of care.

SOCIAL SECURITY AMENDMENTS OF 1983: COST CUTTING INCREASES THE NEED FOR LONG-TERM CARE AS WELL AS CONSUMER ACTIVISM TO IMPROVE QUALITY

Rising health care costs and a fiscally conservative administration in the White House combined in the 1980s to change the federal approach to health care finance in ways that would affect the consumption of long-term care (Kronenfeld, 1997). The Social Security Amendments of 1983, introduced a revised reimbursement schema for hospital care charged to Medicare, based on predetermined rates for diagnostic related groups (DRG's). The result of this financing mechanism was shortened hospital stays for Medicare recipients and a growing reliance on outpatient care by hospitals that would slowly affect the public at large (Wolinsky, 1990). Those in need of assistance after an inpatient episode were forced to rely on family caregivers and extended care facilities where available (Estes, Swan & Assoc., 1993). Since Medicare would cover only a limited number of days of post hospital care, state Medicaid budgets strained to supplement some of this longer-term care. The fact that Medicaid funding to the states had been cut by TEFRA (Tax Equity and Fiscal Responsibility Act) in 1982 (Kronenfeld, 1997), made this a challenging prospect for even the wealthiest of states.[10]

Nursing home bed growth across the country stabilized during the 1980s despite rising demand from the aging population. Illinois mirrored the national average with a small decline (−0.1%) in the number of new beds per 1000 elderly between 1981–1993 (Harrington et al., 1997). However, substantial long-term care investment prior to the 1980s left most of the state, including Chicago, with a relatively healthy supply of nursing home beds (DuNah et al., 1995). Other cities, however, that did not experience the same growth of nursing home beds in African American neighborhoods after Medicaid's inception found that state and federal efforts to control Medicaid expenditures during the 1980s reinforced already existing patterns of racial inequity in access to long-term care. In Pennsylvania, a moratorium on nursing home construction was imposed at a time when long-term care investment was needed in the inner city of Philadelphia (Smith, 1999).[11]

The demand for long-term care continued to increase, however, and consumer organizations became increasingly concerned about the quality and variety of long-term care available. The National Citizens' Coalition for Nursing Home Reform (NCCNHR), joined with other professional and provider organizations to fight for national standards of nursing home care that could be monitored with a uniform assessment tool (Harrington, 1997). Such recommendations had been proposed by an Institute of Medicine (1986) report on long term care. The Nursing Home Reform Act was passed the following year as part of the Omnibus Budget Reconciliation Act of 1987. The legislation required periodic assessment of nursing residents and the development of outcome measures that related directly to patient health. OBRA regulations prohibit the inappropriate use of restraints and mandate the use of a resident assessment instrument for the development of a minimum data set that will allow ongoing review of facility quality (Harrington, 1991). This important law has been credited with a decline in the use of restraints since its passage and improvement of certain outcome measures (Vladek, 1995). The ability of the long-term care consumer organization to organize a broad-based coalition to achieve major reform from a fiscally conservative Congress was recognized by policy makers and the media.[12] Scholars and consumer groups continue to advocate for the improvement of the MDS and nursing home staffing levels (Kane, 1998; IOM, 2000).

Another coalition of long-term care consumers that included younger disabled persons successfully fought legislative proposals to eliminate Medicaid as an entitlement program and institute block grants to the states in order to limit federal expenditures (Estes et al., 2001). The passage of the Americans with Disabilities Act in 1990 empowered persons with disabilities by recognizing their equal rights to work and play much as the Civil Rights Act of 1964 forbid discrimination against women and minorities (Albrecht, 1992). The emerging coalition between the old and the young will push states away from nursing home care in favor of less institutional long-term care options (Kane et al., 1998). A recent Supreme Court ruling, called the Olmstead decision, further directed states to provide the least restrictive environment for persons with disabilities (Niefield et al., 1999).

In the coming decade Medicaid, which improved access to long-term care for low income consumers, will alternate on the legislative agenda between expansion to meet the growing need for coverage of the uninsured and funding cuts to address the budget constraints faced by many states. An increasingly organized and broad-based consumer movement will continue to fight to maintain standards won in recent decades and prevent persistent threats of deregulation of the nursing home industry. As Albrecht (1992) has noted, involvement of low income consumers of color would strengthen advocacy organizations and ensure that future long-term care policies are designed with the needs of all ages and ethnic groups in mind.

CHANGING LONG-TERM CARE CONSUMERS AND THE FUTURE OF FEDERAL POLICY

Throughout the first half of the 20th century, the typical nursing home resident was an older White woman who resided in a home for the aged established to care for elders of her ethnic background although after 1935 she was just as likely to reside in a small proprietary home in a converted mansion. By the second half of the century, federal policy had begun to broaden the base of consumers to whom providers could market. After the passage of Medicaid, African American persons would slowly begin to increase their utilization of nursing home care until the face of the typical consumer had changed.

This transformation could be attributed to increasing numbers of disabled elders of color. While currently, 87% of elderly persons in the U.S. are White (non hispanic), this percentage is expected to decline to 67% by the year 2050 when, it is estimated, 10% will be African American and 16% Latino (U.S. Bureau of the Census, 1996). In addition, rising rates of disability in the U.S. population disproportionately affect African American and low income persons (Albrecht, 1992; Ford, 1992). These trends are likely to effect the proportion of the elderly who utilize nursing home care.

However, disparities among cities suggest that urban market forces, such as racial segregation and historic ethnic patterns in nursing home development, affect nursing home supply and utilization, particularly because African American elders tend to be concentrated in cities. In Chicago, Medicaid seems to have stimulated long-term care investment in low-income communities that might otherwise have remained underserved. While racial and economic segregation drove other health care providers from the poorest communities of the City primarily custodial care remained profitable as long as staffing levels could be kept relatively low. This was possible in Chicago because private paying patients who would demand such amenities had also left neighborhoods where poverty is concentrated (Wilson, 1987).

The preservation of Medicaid is therefore essential to sustain continued access to long-term care for those in need of institutional care but without the range of options available to those who can purchase it. Given that nursing homes in Chicago and other U.S. cities remain almost as segregated as the neighborhoods in which they are located, free market al.ternatives to Medicaid would leave African American elders without nursing home care. Wallace et al. (1998) estimate that the devolution of Medicaid to a block grant would cause a 50% decrease in Black nursing home use (20% among Whites). Unlike universal long-term care insurance, Medicaid has a limited base of support because only those who qualify by virtue of poverty are eligible. The fact that 20% of nursing home residents "spend down" to become eligible for Medicaid (Niefield et al., 1999) has broadened the consumer base and therefore support for its survival as an entitlement program.

Racial segregation makes integrated consumer organizations difficult to organize and sustain. Consumer organizations with a national focus tend to be white and middle class (Albrecht, 1992). In cities, African Americans fight for quality care for their family members through the ombudsman program and in their communities through AOA funded agencies and community centers. Seeking out the support of such urban organizations may further strengthen the growing consumer movement that has formed coalitions between the old and the young.

Long-term care policy makers may consider changing consumer demographics when designing effective programs. For instance, private long-term care insurance has been promoted as a means of improving access to care and reducing state and federal expenditures. Long-term care insurance will serve to reduce the out of pocket expenditures of middle to high income elders and lessen the numbers of families that must be impoverished to qualify for Medicaid support of nursing home care. However, analysis of leading proposals finds that long-term care insurance will not be affordable for the poor even with tax incentives (Niefield et al., 1999). Therefore, such policies risk further narrowing the base of support for those who must continue to rely upon Medicaid reimbursed care.

Persons of all races, ethnic backgrounds and incomes desire access to less institutional long-term care. Illinois is encouraging Medicaid providers to invest in Assisted Living facilities and this state has one of the largest Medicaid waiver programs for community based home care.[13] However, Medicaid was originally designed to support nursing home care, which is still the only long-term care service that states are required to provide under Title 19 of the Social Security Act (Kane et al., 1998). In the coming years, as alternatives to nursing home care are developed, states must consider the equitable distribution of such beds so that the poor are not left with fewer options in their communities than the private pay, long-term care insured person. The expansion of funding for home and community based care under Medicaid would help ensure that the poor have equal access to nursing home alternatives.

CONCLUSION

The passage of Medicaid in 1965 altered the supply of nursing home beds to Chicago's African American community providing formal services that were unlikely to be supplied by the free market. Whether the city's high level of racial segregation was a factor by allowing Medicaid entrepreneurs to lower staffing levels and achieve an economy of scale is unclear. Other segregated cities still experience shortages of Medicaid beds to low income consumers. However, the rising rate of African American utilization of nursing home care nationally suggests that Medicaid has had a significant effect upon supply despite persistent patterns of racial segregation.

Long-term care coverage for the poor must be preserved at the risk of reducing the supply of nursing home beds to segregated communities. The desegregation of American cities would dramatically improve the health and welfare of low income African Americans (Anderson & Massey, 2001) and improve health indicators for society at large (Wilkinson, 1996). However, if cities maintain current residential patterns designers of federal policy would do well to anticipate the effect on the distribution of health care funding. Spatial analysis can support the equitable allocation of government resources to ensure that long-term care consumers of all races and incomes have a variety of high quality services available to them in the coming century.

ACKNOWLEDGMENTS

The author wishes to thank DePaul University's University Research Council for a Competitive Research Grant that helped support this work, Grace Budrys for her thoughtful comments and Aleksandra Wlodarska for her stellar research assistance. An earlier draft of this paper was presented at the American Sociological Association meetings in Washington, D. C., August 2000.

NOTES

1. Medicare covers the first 100 days of nursing home care and some home health care. Medicaid reimburses nearly half of all nursing home care. In addition, Medicaid reimburses some states for personal care and others for home and community based care to those who are eligible by virtue of their income and level of disability. Seven percent of all long-term care costs are currently covered by privately owned long-term care insurance (Long & Liska, 1998; Rice, 2001).

2. Although African American communities in Chicago do not appear to be underbedded, it can not be assumed that there is an adequate supply. Persons of color

have higher rates of functional limitation than White elders (U.S. Bureau of the Census, 1996). In addition, there is evidence that disabled people of color are younger, which would make the traditional supply variable, beds per elderly, a less appropriate measure of unmet need than for White communities (Reed et al., 2001).

3. Based on the poverty literature, a census tract was considered an area of concentrated poverty if 40% or more of the residents were at the poverty level (Jargowsky, 1997).

4. Life expectancy at birth was 47.3 in 1900; 59.3 in 1930 (Gornick et al., 1996).

5. The Social Security Act of 1935 was a broad government assistance program that instituted welfare for dependent children as well as federal grants to the states for elderly adults and blind persons, maternal and child health services and more. In 1956, disability insurance was granted to those permanently disabled with subsequent amendments adding Medicare coverage (Albrecht, 1992).

6. Hospitals that offer a range of inpatient services are included in this analysis while those that offer strictly rehabilitation and psychiatric care are excluded.

7. The 1979 amendment to the Health Planning and Development Act went farther in encouraging institutions to share expensive equipment like CAT scans and specifically requiring the involvement of certain consumer interest groups (Budrys, 1986).

8. This policy, called "rebasing" was discontinued by Medicaid in the 1980s (Goldberg, 1995).

9. Fossett and Peterson (1989) showed that, in a segregated market, health care providers are unlikely to mix Medicaid patients with those who are privately insured.

10. Like most states, Illinois tried to limit Medicaid expenditures by keeping reimbursement rates low but by 1992 the state owed $500 million in unpaid Medicaid bills (Joseph, 1992). In 1981, the Boren amendment had required states to pay a "reasonable" rate to nursing home providers and Illinois was sued by nursing home owners for violating this law. In 1992, the Medicaid Revenue Assessment Act of Illinois was passed to allow the state to "tax" providers, get reimbursement from Medicaid for that money and then give it back to the owners in the form of a rate increase. This scheme has been challenged by HCFA but similar arrangements were devised by other states to obtain a greater level of Medicaid support from Washington (Illinois Department of Public Aid, 1992).

11. In Philadelphia only board and care homes were willing to admit African American elders, while long-term care investors developed facilities in areas of the City where private pay patients lived (Smith, 1999).

12. This victory may be one reason why another coalition of older adults won the repeal of the Medicare Catastrophic Coverage Act of 1988 one year later.

13. Although Illinois' Community Care Program is large, the state relies heavily upon nursing home care (Newcomer, Tonner, LeBlanc, Crawford, Wellin & Harrington, 2000).

REFERENCES

Albrecht, G. L. (1992). *The disability business: Rehabilitation in America.* Sage Library of Social Research 190. Newbury Park, California: Sage Publications, Inc.
Anderson, E., & Massey, D. S. (2001). *Problem of the Century: Racial stratification in the United States at Century's end.* New York: Russell Sage Foundation.

Angel J., Angel R. J., McClellan J. L., & Markides K. S. (1996). Nativity, declining health and preferences in living arrangement among elderly Mexican Americans: Implications for long-term care. *Gerontologist, 36*(4), 464–473.

Angel, R. J. & Angel, J. L. (1997). *Who will care for us?: Aging and long-term care in multicultural America*. New York: New York University Press.

Budrys, G. (1986). *Planning for the nation's health*. Westport, Connecticut: Greenwood Press, Inc.

Clark, D. O. (1997). U.S. trends in disability and institutionalization among older Blacks and Whites. *American Journal of Public Health, 87*(3), 438–440.

Cohen, S. & Scull, A. (Eds) (1983). *Social control and the state: Comparative and historical essays*. Oxford: Martin Robertson.

Cutler, I. (1982). *Chicago: Metropolis of the mid continent* (3rd ed.). Dubuque: Kendall Hunt.

Damron-Rodriguez, J., Wallace, S. P., & Kington, R. (1994). Service utilization and minority elderly: Appropriateness, accessibility and acceptability. *Gerontology and Geriatric Education, 15*(1), 45–64.

Douglas, R., Espino E., Meyers, M., McLelland, S., & Haller, K. (1988). Representation of the Black elderly in Detroit metropolitan nursing homes. *Journal of the National Medical Association, 80*(3), 283–288.

Drake, S. C., & Cayton, H. R. (1945). *Black metropolis: A study of Negro life in a northern city*. New York: Harper and Row.

DuNah, R., Harrington, C., Bedney, B., & Carrillo, H. (1995). Variations and trends in state nursing facility capacity: 1978–1993. *Health Care Financing Review, 17*(1), 183–199.

Estes, C. L., Swan, J. H., & Assoc. (1993). *The long term care crisis: Elders trapped in the no-care zone*. Newbury Park, California: Sage.

Estes, C. L., Weiner, J. M., Goldberg, S. C., & Goldenson, S. M. (2001). The politics of long-term care reform under the Clinton health plan: Lessons for the future. In: C. L. Estes & P. R. Lee (Eds), *The Nation's Health* (6th ed., pp. 206–214). Sudbury, Massachusetts: Jones and Bartlett Publishers.

Ettner, S. (1993). Do elderly Medicaid patients experience reduced access to nursing home care? *Journal of Health Economics, 11*, 293–280.

Falcone, D., & Broyles, R. (1994). Race as a barrier. *Journal of Health Politics, Policy and Law 19*(3), 583–595.

Ford, A. B., Haug, M. R., Roy, A. W., Jones, P. K., & Folmar, S. J. (1992). New cohorts of urban elders: Are they in trouble? *Journal of Gerontolog, 47*(6), S297–S303.

Fossett, J. W. & Peterson, J. A. (1989). Physician supply and Medicaid participation: The causes of market failure. *Medical Care, 27*(4), 386–396.

Ginsberg, E., Berliner, H. S., & Ostow, M. (1993). *Changing U.S. health care: A study of four metropolitan areas*. Westview Press: Boulder.

Gintzig, L., Green, H. A., Reeves, P., Levy, V., Wilcox, J., & Crane, S. (1969). *The evolution of long-term care in the U.S.* Research Division, Department of Health Care Administration, George Washington University, Monograph No. 1.

Glick, F. Z. (1930). Admissions to homes for the aged with special reference to Chicago. Dissertation for Masters Degree, University of Chicago, Chicago.

Goldberg, S. L. (1995). Where have nursing homes been? Where are they going? *Generations*, (Winter), 78–81.

Gornick, M. E., Warren, J. L., Eggers, P. W., Lubitz, J. D., DeLew, N., Davis, M. H., & Cooper, B. S. (1996). Thirty years of Medicare: Impact on the covered population. *Health Care Financing Review, 18*(2), 179–237.

Harrington, C. (1991). The nursing home industry: A structural analysis. In: M. M. Minkler &
    C. L. Estes (Eds), *Critical Perspectives on Aging: The Political and Moral Economy of
    Growing Old* (pp. 153–164). Amityville, New York: Baywood Publishing Company, Inc.
Harrington, C. (1997). Nurse staffing: Developing a political action agenda for change. In: C. L.
    Estes & P. R. Lee (Eds), *The Nation's Health* (5th ed., pp. 259–269). Sudbury,
    Massachusetts: Jones and Bartlett Publishers.
Harrington, C., Swan, J. H., Nyman, J. A., & Carrillo, H. (1997). The effect of certificate of need and
    moratoria policy on change in nursing home beds in the United States. *Medical Care, 35*(6),
    574–588.
Hospital Planning Council for Metropolitan Chicago (1968). *Utilization and status of nursing
    homes and nursing care units in homes for the aged in the Chicago metropolitan area
    for calendar years 1964–1966.* Chicago, Illinois: Hospital Planning Council for Metropolitan
    Chicago.
Illinois Department of Public Aid (1992). Financing Medicaid – Provider assessment. Paper prepared
    for the Governor's Health Care Reform Task Force. Chicago, Illinois: Illinois Department of
    Public Aid.
Institute of Medicine. (1986). *Improving the quality of care in nursing homes.* Washington, D.C.:
    National Academy Press.
Institute of Medicine. (2000). *Improving the quality of long-term care.* Washington, D.C.: National
    Academy Press.
Jargowsky, P. A. (1997). *Poverty and place: Ghettos, barrios and the American city.* New York, NY:
    Russell Sage Foundation.
Joseph, L. B. (Ed.) (1992). *Paying for health care: Public choices for Illinois.* A Chicago Assembly
    Book, University of Chicago: University of Illinois Press.
Kane, R. L. (1998). Assuring quality in nursing home care. *Journal of the American Geriatrics Society,
    46,* 232–237.
Kane, R. A., Kane, R. L. & Ladd, R. C. (1998). *The heart of long-term care.* New York: Oxford
    University Press.
Kasarda, J. D. (1993). Cities as places where people live and work: Urban change and neighborhood
    distress. In: H. G. Cisneros (Ed.), *Interwoven Destinies: Cities and the Nation.* New York:
    W. W. Norton and Co.
Katz, M. B. (1986). *In the shadow of the poorhouse: A social history of welfare in the U.S.* New York:
    Basic Books.
Knupfer, A. M. (1996). *Toward a tenderer humanity and a nobler womanhood: African American
    women's clubs in turn of the Century Chicago.* New York: New York University Press.
Kronenfeld, J. J. (1997). *The changing federal role in U.S. health care policy.* Westport, Connecticut:
    Praeger.
Long, P., & Liska, D. (1998). *State facts: Health needs and Medicaid financing.* The Kaiser
    Commission on the Future of Medicaid.
National Center for Health Statistics (2000). *Health 2000.* Hyattsville, Md.: National Center for Health
    Statistics.
Newcomer, R. J., Tonner, M. C., LeBlanc, A. J., Crawford, C., Wellin, V., & Harrington, C. (2000).
    Medicaid home and community based long-term care in Illinois. San Francisco, CA:
    University of California, San Francisco.
Niefield, M., O'Brien, E., & Feder, J. (1999). Long-term care: Medicaid's role and challenges. Kaiser
    Commission on Medicaid and the Uninsured, Policy Brief. Henry J. Kaiser Foundation.
Nursing homes sue state (1979, December 20) *Chicago Tribune, 8.*
Nursing homes win ruling (1980, May 20) *Chicago Tribune, 13.*

Nyman, J. A. (1988). Excess demand, the percentage of Medicaid recipients and the quality of nursing home care. *Journal of Human Resources, 23*(1), 76–92.

Nyman, J. A. (1993). Testing for excess demand in nursing home markets. *Medical Care, 31*(8), 680–693.

O'Connor, A., Tilly, C., & Bobo, L. D. (2001). *Urban inequality: Evidence from four cities.* New York: Russell Sage Foundation.

Reed, S. C., & Andes, S. (2001). Supply and segregation of nursing home beds in Chicago communities. *Ethnicity and Health, 6*(1), 35–40.

Reed, S. C., Andes, S., & Tobias, R. A. (2001). Concentrated poverty and the distribution of nursing home care in Chicago. *Journal of Health Care for the Poor and Underserved 12*(1), 88–102.

Rice, D. P. (2001). Medicare: A women's issue. In: C. L. Estes & P. R. Lee (Eds), *The Nation's Health* (6th ed., pp. 514–522). Sudbury, Massachusetts: Jones and Bartlett Publishers.

Richardson, B. K. (1963). *A history of the Illinois Department of Public Health, 1927–1962.* Springfield, Illinois: State of Illinois: Rothman.

Rothman, D. J. (1971). *The discovery of the asylum.* Boston: Little Brown.

Rowland, D., Feder, J, Lyons, B., & Salganicoff, A. (1992) *Medicaid at the crossroads.* Kaiser Commission on the Future of Medicaid, Baltimore, Maryland: Henry J. Kaiser Family Foundation.

Salmon, J. W. (1993). Chicago health care: Private growth amidst public stagnation. In: E. Ginsberg, H. S. Berliner & M. Ostow (Eds), *Changing U.S. Health Care: A study of four metropolitan areas* (pp. 41–97). Boulder: Westview Press.

Scanlon, W. J. (1980). A theory of the nursing home market. *Inquiry, 17,* 25–41.

Skocpol, T. (1992). *Protecting soldiers and mothers: The political origins of social policy in the United States.* Cambridge: Belknap Press.

Smith, D. B. (1999). *Health care divided: Race and healing a nation.* Ann Arbor: University of Michigan Press.

Spear, A. H. (1968). *Black Chicago: The making of a Negro ghetto, 1890–1920.* Chicago: University of Chicago Press.

Squires, G. D., Bennett, L., McCourt, K., & Nyden, P. (1987). *Chicago: Race, class and the response to urban decline.* Philadelphia: Temple University Press.

U.S. Bureau of the Census. (1996). Current Population Reports. Special Studies P23–190. *65+ in the United States.* U.S. Government Printing Office. Washington, D.C.

Vladek, B. C. (1980). *Unloving care: The nursing home tragedy.* New York: Basic Books.

Vladek, B. C., & Feuerberg, M. (1995). Unloving care revisited. *Generations,* (Winter), 9–13.

Waitzman, N. J., & Smith, K. R. (1998). Separate but lethal: The effects of economic segregation on mortality in metropolitan America. *Milbank Quarterly, 76*(3), 341.

Wallace, S. P. (1990). Race versus class in the health care of African American elderly. *Social Problems, 37*(4), 517–534.

Wallace, S. P. (1993). Nursing home segregation in major metropolitan areas. Paper presented at the Society for the Study of Social Problems Annual Meeting, Miami Beach.

Wallace, S. P., Enriquez-Haass, V., & Markides, K. (1998). The consequences of color-blind health policy for older racial and ethnic minorities. *Stanford Law and Policy Review, 9*(2), 329–346.

Whiteis, D. G. (1992). Hospital and community characteristics in closures of urban hospitals, 1980–1987. *Public Health Reports, 107,* 409–416.

Wilkinson, R. (1996). *Unhealthy societies: The afflictions of inequality.* London: Routledge.

Wilson, W. J. (1987). *The truly disadvantaged: The inner city, the underclass, and public policy.* Chicago: University of Chicago Press.

Wittenborn, E. L. (1983). *A history of the Illinois Department of Public Health: 1962–1977.* Springfield: State of Illinois.

Wolinsky, F. D. (1990). *Health and health behavior among elderly Americans: An age-stratification perspective.* New York: Springer Publishing Company.

Zinn, J. S. (1994). Market competition & the quality of nursing home care. *Journal of Health Politics, Policy and Law, 19*(3), 555–582.

Zinn, L. (1999). A good look back over our shoulders. *Nursing Homes Long Term Care Management, 48*(12), 20–39.

COMPUTER TOOLS AND SHARED DECISION MAKING: PATIENT PERSPECTIVES OF "KNOWLEDGE COUPLING" IN PRIMARY CARE

Robert R. Weaver

ABSTRACT

Two parallel developments in health care have begun to converge: (1) the demand for greater patient participation in health care, and (2) the evolution of computer tools designed to inform patients of health care options so they might decide among them. Little is known about how patients see these tools and how these perceptions affect participation in care. This study examines patients' responses to the routine use of "knowledge coupling" computer tools in primary care, and their influence on "shared decision making". Three hundred fifty seven (357) patients responded to a survey that asked about various aspects of the knowledge coupling system used in a primary care practice. The results indicate that how patients see the knowledge coupling tools used in the practice affects perceptions of their involvement in care and decisions. Perceptions of involvement also positively relate to age, but remain unrelated to gender, education, income, and home computer use. Trends toward using computer tools to assist in health decisions will likely continue. This study suggests that as these tools become integrated into the

Changing Consumers and Changing Technology in Health Care and Health Care Delivery,
Volume 19, pages 131–149.

routine care of patients they will have important implications for patient
participation in care and decision making.

INTRODUCTION

Recently, two separate and parallel developments in health care have begun to converge. The first is a trend toward greater patient demand for information about and control over their own health and health care. The second is the evolution of computer tools designed to help provide patients (and caregivers) with knowledge and information specifically suited to their unique health situation. This convergence can dramatically alter how medical knowledge and information is processed and shared, and so changes the nature and structure of caregiver-patient interaction. To date, we know little about whether and how new computer tools will be welcomed by or unsettling to the patients they affect, or whether they will "empower" or "disempower" them to take charge of their own health (Hersey et al., 1997). The reasons why computer tools for routine care have not seen wide use yet are numerous and beyond this chapter's scope (Weaver, 1986, 1991; Morris, 2000; Christiansen et al., 2000). Suffice it to say they evoke both fears and hopes for the future of health care (Maxmen, 1976; Lesse, 1983, 1986; Reiser, 1978; Ritzer, 2000).

There are, however, instances wherein computer tools have been used in a thoroughgoing way in the routine care of patients, and which are designed to enable patients to play the primary role in the care process. This paper is about perhaps the most "radical" approach to the use of computers in health care, embodied in tools called "knowledge couplers" (Weed, 1991, 1994, 1995, 1997, 2000). The bulk of this chapter is about the use of knowledge coupling (KC) tools in a primary care practice (PCP), patients' perceptions of these tools, and how these perceptions influence active participation in "shared decision making" (SDM). Because this approach is so radical in design and is used for routine care, its investigation offers a possible window into how this and other computer tools might further the transformation of the caregiver-patient relationship in the future. First, however, I briefly review changes in the physician-patient relationship that have been ongoing for over two decades, and then discuss why the use of computers is necessary to ensure that up-to-date knowledge is routinely applied in the everyday care of patients.

NEW CONSUMERS OF HEALTH CARE

The traditional "paternalist" model of doctor-patient interaction assumes that modern medical care involves dispensing medical advice and deploying technical skills that, by virtue of their lengthy training and technical competence, physicians are best qualified to do (Parsons, 1951). This model further assumes that the physician plays the dominant role in the relationship with the patient, serves as the guardian of the patient, knows what is best for the patient, and decides and acts in the patient's best interest (Emmanuel & Emmanuel, 1992; Charles et al., 1999).

The paternalist model has faced serious challenges in recent decades (Charles et al., 1997, 1999; Stevenson et al., 2000). Medicine and medical decision making are increasingly understood to involve human values, instead of being strictly a matter of applying the best technique. For instance, accumulating evidence shows significant variation in decision making, depending on the practice setting, practice styles, or patient characteristic (Eddy, 1984; Eisenberg, 1986; McKinlay et al., 1996; Payer, 1988; Wennberg, 1984). This suggests how various non-technical factors influence physicians' decisions, that these decisions are not value neutral, and their values are not always identical to those of their patients. Further, with medical advance a broader range of treatment options come available to manage a single condition, each with its own benefits and costs. Weighing these costs and benefits entails deciding among values. Value decisions, in contrast to strictly technical ones, are perhaps most appropriately made by patients who, in the end, must live with their consequences (Eddy, 1990; Levine et al., 1992). Even the alternative of "watchful waiting" offers an option that patients with various conditions might have good reason to consider. More and more patients see it as their business to be informed about and to decide among treatment alternatives. These changes coincide with a growing tendency to challenge the physician's authority (Haug, 1973; Starr, 1982; McKinlay & Arches, 1985; Leyerle, 1984). This leaves physicians more vulnerable to questioning than before, especially among a more skeptical and discerning consumer of health care. Though a weak guarantor of genuine patient participation (providing patients with little more than "veto power" in decision making), the legal demand for "informed consent" represents the most obvious manifestation of the larger movement toward greater patient participation in health care (Appelbaum et al., 1987).

Increasingly inquisitive patients demand more knowledge and information about their conditions, the various treatments available to manage them, and the tradeoffs associated with each. Patients so informed become "empowered" to make decisions about what management strategy is most suitable to their needs and values, or at least share in this decision. Hence, the traditional and longstanding paternalist model is fast becoming supplanted by alternatives – variously understood as "informative" or "shared" patient care models (Emanuel & Emanuel, 1992; Charles et al., 1999). For decision making authority to have meaning, patients must first be *informed* about their condition and the various treatment alternatives available to manage it. This demands that much more information must be exchanged between the caregiver and the patient to enable the patient to meaningfully participate in health decision making. Not only do patients want their complaints to be "heard and understood" by the physician, paramount for them is to "know and understand" what it is that is ailing them (Ong et al., 1995; Deber et al., 1996; Cegala, 1997). Even if patients ultimately "decide not to decide" or, rather, to finally defer to the physician's decision, most prefer to be as informed as possibly about their condition (Nease & Brooks, 1995; Guadagnoli & Ward, 1998).

Though it still persists to a great degree in everyday medical practice and is resistant to change, ample evidence suggests that the traditional paternalist model is on the decline and that alternative "shared" or "informative" models are taking its place. As discussed, the latter presupposes that patients are fully informed about their condition, regardless of the actual role they play in decision making. Even well-intended physicians who wish their patients to be fully informed fall short of this ideal, either due to the lack of time or inadequate interpersonal skills (Stevenson et al., 2000). Even more critical is that *caregivers* themselves are unable to reliably recall relevant information or to be fully informed about all available options while discussing matters with patients. In view of the vast and changing body of medical knowledge, it is hard to believe that one could ever be. As discussed below, computer systems that enable rapid and reliable access to knowledge and information, and which can organize that information in useable ways, are needed to overcome this wholly human limitation.

WHY COMPUTER TOOLS ARE NEEDED IN MEDICAL CARE

Leaving aside for the moment the issue of how values and interests are implied in each medical decision and whether and to what extent the physician and the patient share the same values and interests, what assurance is there that all the various options from which to choose are even presented for consideration? There is, in fact, ample evidence to suggest that the range of options available to patients is not and cannot be presented to patients without the help of external, computer aids. First, and most obviously, providers cannot keep up with the vast and ever-expanding body of medical knowledge. It is impossible for anyone to reliably consider the most up-to-date knowledge about the diagnostic and management options when solving patients' problems (Covell et al., 1985; Hunt & Newman, 1997). Second, studies in psychology have for years detailed numerous flaws and biases that contaminate the problem-solving process – e.g. formulating diagnostic hypotheses prematurely, seeking out data that confirms rather than disconfirms the hypotheses, misapplying population-based information to individual cases (Tversky & Kahneman, 1974; Elstein, 1976; Voytovich et al., 1985; Dawes et al., 1989; Weed, 1991). As mentioned above, others have shown how a range of non-medical factors – e.g. characteristics of patients, physicians, practice settings, and the culture at large – influence medical decision making (Eisenberg, 1986; McKinlay et al., 1996; Payer, 1988). Flaws in decision making are compounded, of course, as providers face different patients, some with multiple problems, every 10 minutes or so. In view of this lack of control over the cognitive inputs to decision making, it is no surprise that we find high variation in medical practice (Wennberg et al., 1984; Eddy, 1984), inappropriate use of medical resources (Eisenberg, 1986) and high amounts of preventable medical error (Leape, 1994).

As computers offer caregivers access to medical knowledge and information they can make this knowledge and information available to patients as well. Inasmuch as patients are included in the *process* of collecting and organizing relevant knowledge and information, they are in a better position to meaningfully participate in the decision making at the end of the process. Properly deployed, computers can elevate the extent of knowledge and information exchange and help satisfy patients' need to be understood and to understand their own condition. In so doing, the tools can also bolster patients' trust and confidence in the care they receive, thereby helping to satisfy emotional needs that are also important for patients (Hall et al., 1987; Roter et al., 1987; Ong et al., 1995; Cegala et al., 1996; Cegala, 1997; Roter et al., 1997; Roberts & Aruguete, 2000). Moreover, these aids enable knowledge and information to be delivered to patients in formats that are specific and suitable to them – instead of receiving only oral instructions from the physician. This gives patients time to contemplate and discuss the options available and to make an informed choice about which one best suits their needs. Hence, it is likely that information tools will increasingly find a place in the routine care of patients with the potential to elevate the level of knowledge and information exchange, while they can also serve to increase patients' confidence in and control over the health care process.

To recap, recent decades have seen the convergence of two parallel trends in medicine: (1) greater patient participation in medical care decision making, and (2) the development of computer systems to facilitate access to knowledge and information needed to make decisions. Greater patient participation requires greater exchange of information between providers and patients and takes the form of "shared" or "informed" models of medical care and decision making.

Both patients *and* providers need external aids (computer tools) to facilitate access to relevant medical knowledge and information, to reduce medical uncertainty, and to elevate confidence in medical care. This paper examines the use of one such computer system called "knowledge coupling" that is designed to facilitate routine access to medical knowledge and information and to enable patients to play a greater role in medical decisions.

"KNOWLEDGE COUPLING" COMPUTER TOOLS AND NEW CONSUMERS OF HEALTH CARE

Knowledge Coupling Tools to Guide Decisions

Various decision aids are now available to help inform patients about their health situation or even to empower them to make decisions. These "informatics tools" include simple brochures that contain generic information about diseases, fact sheets, interactive video discs, video and audio tapes, and computer systems for direct patient use or to generate fact sheets tailored to each individual patient. To date, evidence for the successful use of decision aids to enable patients to meaningfully participate in decisions concerning a range of health conditions – e.g. adult immunization (Carter et al., 1986), benign prostatic hyperplasia (Flood et al., 1996), quality of life for patients with H. I. V. (Gustafson et al., 1998) – while limited, remains promising (Hersey et al., 1997).

Most informatics tools are used to *provide* patients with information in various formats at the very end of the problem solving process. Seldom are they used to elicit and offer information and knowledge throughout the health care process. In contrast, knowledge coupling (KC) tools are designed to guide patients *and* caregivers in the collection and management of medical information during the process of problem solving (Weed, 1991, 1994, 1997, 2000). In addition to providing information to patients about the (already delimited) choices available, the tools serve as a key "listening" device used to gather information about patients' complaints and to align the information with pertinent medical knowledge. Central to the KC paradigm (from which the tools are an outgrowth) is that patients actively participate in and contribute to the problem solving process; ultimately, patients are the primary decision makers.

How do KC tools help guide the process of collecting information used to inform medical decisions? First, they identify all the known causes of a particular problem and a set of information to collect for discriminating among these causes. For instance, the adult medical *history* questionnaire includes over 700 items, while the questionnaire for *diagnosing* depression includes over 400. Responses to a similarly broad ranging set of questions are also used as input to inform decisions regarding the *management* of complex problems such as, for instance, diabetes. Hence, the thoroughness with which patient problems are explored exceeds what is possible without computer aids. Once the information is collected and input, the tools align or "couple" these patient-specific findings with related medical knowledge and present the range of diagnostic and management options as they relate to the patient's unique set of findings. The patient, the caregiver, or both then interpret the results of the coupling process. Moreover, the information collected and decisions made become part of the patient's medical record that can be shared with the patient.

KC Tools and Patients

If implemented as Weed envisions, KC tools will enhance the amount of knowledge and information exchanged between the caregiver and the patient, and, thus, help satisfy the patient's need to "be heard and understood", and to "know and understand" what is ailing them. Patients will not only be more involved throughout the problem solving process, in the end they will possess the medical knowledge and information relevant to their unique situation. In this view, since patients maintain their own set of values and preferences, they are best suited to weigh the pros and cons of various options, and to make the medical decision most appropriate for their particular situation. Even if they decide to defer to the caregiver's judgment on this matter, they can do so with the knowledge that their case has been thoroughly explored and that all the relevant diagnostic and management options have been considered.

The Setting: KC Tools in a Primary Care Practice

How have KC tools been integrated into the daily routine of the primary care practice (PCP)? In brief, before the actual office visit, the PCP distributes problem-specific questionnaires to patients complaining of a problem. The questionnaires take the form of a checklist designed to elicit an array of information about the nature and progress of the problem; patients merely check an item to indicate the presence of a particular finding. The PCP receives the questionnaire before the visit and inputs the set of positive findings that the patient identified as characterizing her or his particular condition.

During the actual office visit a medical assistant verifies these findings with the patient and collects additional information about the patient's physical condition – information that patients cannot provide for themselves. Information from the questionnaire and physical examination serves to define the patient's condition. Once the findings have been collected, input, and reviewed, they are "coupled" with medical knowledge related to them. The results of the coupler session might yield an unambiguous diagnosis and a clearly favored management plan (wherein the diagnostic or management decision makes itself) or two or more plausible options might present themselves. Either way, it is left to both the patient and the provider (i.e. the physician or nurse practitioner) to interpret, discuss, and evaluate these results and to construct a plan to manage the patient's problem. A record of the positive findings, diagnosis, and management plan is then entered in a set of "office notes" that is then printed out and turned over to the patient at the end of the office visit (Burger, 1997).

RESEARCH QUESTION

Based on the above considerations, this study addresses the following research question: How do patients' views of the KC tools affect their perspectives on participation in the "shared decision making" (SDM) process the PCP seeks to promote?

METHOD

Questionnaire

This paper reports the results of a questionnaire distributed to patients in the PCP that uses KC tools routinely in the care of patients. Questions were both open and close-ended and asked how patients viewed various aspects of the KC system used in the practice, along with their overall satisfaction with care (Peersman et al., 2001). Questions about the KC system explored the nature of the information exchange that occurred during the consult as a consequence of using KC tools. Close-ended questions help assess how respondents view the information tools (e.g. the use of the computer and office notes) in relation to their participation in care and decision making. Responses to the open-ended questions enable us to gauge how respondents understand the tools and their uses, their strengths and their limitations.

Sample

Two waves of questionnaires were sent to 661 of 3216 patients (20%) the PCP identified as "active" on its list of patients. Identification numbers attached to the questionnaires enabled us to protect respondents' anonymity so caregivers could not know who might have offered negative (or positive) views about the practice. Of the 661 selected, 357 completed and returned the questionnaire (54%). Of those who responded, 212 (59%) replied to open ended questions; remarks ranged from short sentences (usually about their general impressions) to more lengthy and thoughtful comments about the tools and their use.

Measures

Survey items used a five-point scale (strongly disagree to strongly agree) to assess respondents' orientations toward the computer, office notes, and decision making at the PCP. Negatively worded survey items were reverse coded so all were inclined in the same direction. Items were then subjected to principle factor analysis (using a varimax orthogonal rotation method). Cronbach alpha scores were also computed for items that constitute each factor; scores for items associated with each factor were averaged to create the dependent and intervening variable constructs (see Table 1).

Dependent Variable
Responses to questions that ask about whether respondents act as "equal part-ners" in the PCP and whether they play a "more active role" in the PCP compared to other practices combine to generate a "shared decision making" (SDM) factor.

Intervening Variables
Measures of respondents' orientations toward the computer and toward the use of office notes were constructed to assess the affects, if any, such orientations had on SDM. Factor analysis of responses to questions about the use of the computer generated two factors. The first factor (C-Interf) pertained to the extent to which respondents saw the provider as "relying too much" on the computer and whether the computer "interfered" with communication. The second factor (C-Share) pertained to respondents' more favorable view towards the computer and its contribution toward sharing information. Two additional factors were generated based on responses to questions about the practice of sharing office notes. The first (ON-Understand) pertained to respondents' view of providers' explanation of office notes and their own understanding of them; the second (ON-Use) pertained to their views of the use (outside the office) and importance of office notes.

Socio-demographic Variables
Finally, measures of age, gender, education, family income, and home computer use represent the socio-demographic variables explored in the analysis.

Table 1. Variables Used in the Study.

Variables (Constructs)	Question Items Used
Dependent variable	
SDM (shared decision making) (Alpha = 0.70)	• In this practice I feel as though I am an equal partner in decisions about managing my health. • Compared to other practices, this practice allows me to take a more active role in managing my own health.
Intervening variables	
C-Interf (Computer interferes) (Alpha = 0.88)	• The computer sometimes interferes with the provider's ability to hear my complaints. • The computer does not affect my interaction with the provider one way or another. [reverse coded] • The provider relies too much on the computer to assess my health situation.
C-Share (Computer aids in sharing information) (Alpha = 0.84)	• I see the computer as playing an important role in the care the provider gives. • The computer improves the ability to share information about my health situation. • I do not care much one way or another about the use of the computer in the office. [reverse coded] • I would rather the provider not use the computer in the office. [reverse coded] • I would recommend that other providers use computers to assist in assessing the patient's health situation.
ON-Understand (Office notes enhance understanding) (Alpha = 0.77)	• I understand all or most of the information I receive from the office about my health situation. • The provider explains the office notes I receive so that I understand them. • The provider is open about sharing information with me about my health situation.
ON-Use (Use office notes) (Alpha = 0.80)	• I keep a copy of the office notes handy so I can get to it if and when I need to. • I have had occasion to refer to the office notes for information about my health situation. • I have used the office notes to discuss my health situation with others outside the office (e.g. family, other care providers) • It is important to me that I receive a copy of the office notes before leaving the office.

Table 1. Continued.

	• I would recommend that other providers share the office notes with the patient.

Socio-demographic variables

Age	Respondent's age
Gender	Respondent's gender
Education	Highest educational degree respondent has earned
F-Income (Family income)	Respondent's reported family income
HC-Use (Home computer use)	Whether or not the respondent reports to use a computer at home

RESULTS

Characteristics of Respondents

Fifty-seven percent of the respondents were female, while 43% were male. The average age of the respondents was 52 years, the youngest being 21 and the oldest being 88. Respondents were more "well-to-do" than the general population in terms of education, income, insurance status, and access to computers at home. Only 16 (4.5%) had not received a high school degree or General Education Degree, while 91 (26%) obtained a Bachelor's Degree and another 61 (17%) received a graduate degree. The higher-than-average education also corresponds with higher-than-average income. Forty-five (14%) respondents reported living on an annual family income of less than $20,000, while 68 (21%) earned family incomes of $70,000 or more. The annual family income category at the mid-point of the distribution was $40,000 to $49,999. Most patients (79%) were covered by private insurance only, 18% received insurance through Medicare, and 2.8% were insured through Medicaid. Finally, 153 respondents (43%) indicated that they "use a computer at home" for either e-mail purposes, accessing the World Wide Web, or to obtain health information.

Perception of KC Tools and Shared Decision Making

Model 1 in Table 2 shows the relationships between perceptions of the computer and office notes and shared decision-making (SDM). Model 2 in the same table shows the associations when various socio-demographic variables are included in the regression equation.

Table 2. Regression of Shared Decision Making (SDM) on Perceptions of
Computer Use, Office Notes, and Paternalism, (Model 1) and Patient
Attributes (Model 2).

Variables	Model 1		Model 2	
	b	Beta	B	Beta
C-Interf-	0.001	−0.027	−0.001	−0.002
	(−0.397)		(−0.021)	
C-Share	0.113	0.282**	0.139	0.336**
	(3.789)		(4.349)	
ON-Understand	0.224	0.266**	0.250	0.287**
	(4.569)		(4.517)	
ON-Use	0.091	0.239**	−0.072	0.211**
	(3.816)		(3.126)	
Age			0.014	0.127*
			(2.058)	
Gender			0.137	0.047
			(0.804)	
Education			−0.021	−0.023
			(−0.374)	
F-Income			0.023	0.030
			(0.483)	
HC-Use			0.092	0.031
			(0.478)	
Constant	2.957		1.344	
R^2	0.426		0.436	

* $p < 0.05$; ** $p < 0.01$ (two-tailed test).
Note: The b refers to unstandardized and Beta to standardized coefficients. The t-values appear
in parentheses below the coefficients.

In both models, the perception of the computer as interfering with
communication with the caregiver (C-Interf) shows little affect on SDM, while
the view that the computer enhances information sharing (C-Share) positively
influences the perception of active participation and equal involvement in
decision making. Similarly, respondents who indicate more favorable orientation
toward their use (ON-Use) and understanding (ON-Understand) of office notes
also tend to score higher on the SDM indicator.

Of the socio-demographic variables, only age affects SDM, though in
unexpected ways. While the stereotypic older patient plays a passive role in
patient care, the data from the table suggest otherwise – as age increases, so
does SDM. It might be that older patients, having been in the practice longer,
are more familiar with the caregivers and their practice style, and are more
accustomed to the PCP's use of the computer and office notes. Hence, they

might be more comfortable participating. Other socio-demographic variables – gender, education, income, and home computer use – did not influence respondents' orientation toward more or less SDM. Moreover, the socio-demographic variables showed little affect on perceptions of the computer and office notes. Regression of each of the four intervening variable constructs on the five socio-demographic variables (not shown in Table 2) found only education to be associated with C-Interf (Beta = 0.16, p = 0.012) and age to be associated with ON-Use (Beta = 0.14, p = 0.041).

Written comments – both favorable and unfavorable – help us understand how patients see the computer, office notes, and SDM in the practice. For instance, numerous respondents suggest how the computer presents and helps promote the discussion of various available options, while they, in turn, are given discretion to select the alternative most suitable for them. As one respondent comments:

> I like the computer because it compiles all your complaints and symptoms and gives you all of the possible medical problems that you could have. I think sometimes doctors just send you home with a possible diagnosis, not really knowing if it is more serious . . . I like being able to read a computer screen showing everything!

In contrast, others suggest a more negative orientation toward the computer, seeing it as contributing to an "impersonal" health care climate. Such a view contributes to a sense of alienation, a more passive orientation toward their patient care, and a disinclination to participate as an equal partner in decision making. A few express ambivalence toward the computer, suggesting how beneficial it is for accessing and organizing information, while it also interferes with their dialogue with the caregiver. A few others are less equivocal: "I would prefer that the computer would be eliminated. Again it is just another way that the patient is alienated from the health care provider. Computers are mechanical and impersonal."

In a busy practice the opportunity to discuss *all* the options is largely limited by the time available for each patient – approximately 10–15 minutes per visit. By enabling a broad range of options to be considered, computers in a way promote more lengthy and costly (to the provider) office visits. This problem is, in part, resolved by providing patients with printouts of information associated with their condition. Most patients understand and use office notes and see themselves as more actively involved in the practice and in decision making. Just a handful of respondents offered negative comments about the practice of sharing office notes, suggesting that they are too "impersonal". The overwhelming majority expressed favorable views. Several respondents suggest that the notes enhance "confidence" in the care they receive and encourage their more active involvement: "I feel more confident reading about my information and comparing it to how I feel and what I have for symptoms. It makes me feel reassured about treatment." Another suggests that this reassurance encourages greater involvement: "I like the idea that all of my information is being shared with me. I like to be that much more a part of my own health care." Office notes resemble various other decision tools designed to promote patient decision making, but they directly connect to what occurred during the office visit itself and are tailored to the patient's particular condition. One response illustrates how they serve to encourage patients to involve themselves in their own care:

> [The] office notes in my medical folder . . . remind me of my current health trends, and what I need to do to help myself, if there are medical reasons to do so. Also, they show any progress made and increase my awareness of what I need to focus on. Also, it is comforting to me, to have self-knowledge of my overall health status . . .

While the notes assist patients in understanding their own condition, they also support efforts to explain it to others outside the office. Respondents write about using them to discuss their health condition with family, trusted friends, with other medical professionals, with insurers and employers. "I like being able to share the notes with my spouse. Then she can keep up with what's going on. With the notes I need not remember everything that goes on at my appointment. It is also very good to be able to compare one visit to another at my leisure." This sharing of health information with others, in turn, advances understanding of their condition, and helps activate broader and more informed involvement in their care.

In sum, the results suggest that respondents' perceptions of KC tools influence their views toward their participation in the patient care and decision making process. This influence persists irrespective of age, gender, education, and income, and is not influenced by respondents' computer use in the home. Of the socio-demographic variables examined, only age shows a direct positive affect on SDM.

DISCUSSION

In large part, playing the role of equal partner in health care decisions represents the "tip of the iceberg" when it comes to patient involvement in their health care. The inclusion of patients throughout the entire *process* of patient care is necessary so they might meaningfully participate in decision-making. Some time ago Appelbaum et al. (1987) described the hollow meaning of the "informed consent" received from most patients who do not play an active role throughout the problem solving process. In fact, other studies suggest that patients might well rather be part of the overall process of care – wanting complaints to be heard and understood and to understand what is ailing them and how it might be managed – even if in the end they defer to the caregiver when it comes to making critical decisions (Ong et al., 1995; Nease & Brooks, 1995; Deber et al., 1996; Cegala et al., 1997; Guadagnoli & Ward, 1998). Irrespective of who ultimately decides, recent decades have seen more and more patients demand that they be included throughout the process of their health care. The paternalist model of the knowing and caring physician and the passive and deferent patient is giving way to alternatives wherein patients actively participate in their own care, seek more and more information about their condition, and, if they wish, decide which treatment will be administered.

The evolution of various computerized decision support tools reinforces this tendency toward the inclusion of patients in the process of medical decision making (Hersey et al., 1997). Most tools offer patients knowledge and information in various formats so they can make informed choices among various alternatives. Few, however, have been fully integrated into the routine process of patient care to facilitate patients' ongoing participation and decision making. Unlike most other tools, however, KC tools are explicitly designed to encourage patient involvement in routine care, health care management, and decision making (Weed, 1991, 1994, 1995, 1997, 2000). The alternative paradigm the tools embody directly challenges the traditional "paternalist" model of the physician-patient relationship and seeks to place the patient at the center of the health care process. The KC tools identify the range of options available along with their pros and cons; the KC philosophy sees the choice among these alternatives as a choice among values that the patient is in the best position to make.

How and to what extent this alternative paradigm is actualized in practice and how patients view these tools and their role in their own care is a different matter, however. The array of social constraints that modify how computer tools become introduced are numerous and well beyond the scope of this paper (c.f. Weaver, 1991; Kaplan, 1995; Berg, 1997, 1998). The present study offers evidence indicative of the important role they can play in influencing patients' participation in clinical care and decision making. How patients view the tools and their use affects the extent to which they see themselves as active partic-ipants and equal partners in the health care process.

This study faced certain limitations that point us toward areas that future research would do well to explore. Although the present study showed that respondents' views of the computer and office notes affect the extent to which they perceive themselves as active participants and equal partners in decision making, unexplored are the differences in SDM between patients whose care entails the comprehensive use of computers throughout the process of care (as in the PCP examined here) and patients in more traditional practices that do not use computers during the patient care process. Although in view of evidence from this investigation I suspect that the routine use of the computer and office notes does foster, and to some degree demands, greater patient participation in care, more explicit comparisons are necessary to say so more conclusively.

Relatedly, it is difficult to distinguish, except in the abstract, the user from the tool itself. This is particularly so with KC tools in that they embody a broad philosophy of care developed over several decades, a philosophy that is largely embraced by the caregivers in the PCP. This makes it difficult to assess the relative contribution of the philosophy of care (as expressed in the caregiver's orientation toward patients) and the tools themselves to patient participation. To do so, future research might examine variation in the skill and orientations of the caregivers that use the tools, and how this variation impacts patient participation and involvement in decision making.

For their part, patients no doubt vary in terms of their understanding of KC tools and of the alternative approach to health care the tools imply. The present study found no significant difference in attitudes between patients who used or did not use computers in their home, but this finding remains quite preliminary. Neither did patients' level of education seem to affect their views. I suspect that patients with greater understanding of the "limits of the unaided mind", of how these computer tools are designed to overcome these limits, and of the central role patients can play in health care will view the use of such tools more favorably and will also play a more active role in their health care and in decision making. This remains, of course, a hypothesis that future research can explore.

Sociological studies have already advanced our understanding of the variety of changes that have taken place in the relationship between the physician and the patient. Our understanding of how the integration of computerized decision support tools shapes this relationship is significantly more limited. For the most part this is a consequence of the rather slow introduction of these tools into everyday clinical practice. Increasingly, however, computerized decision support tools are being used to offer patients information about alternative health care options for them to examine, contemplate, and decide among. No doubt this trend will continue, and such tools will become more and more integrated throughout the broader process of patient care as well. The investigation offered herein suggests a great deal of promise that such tools can enhance meaningful patient participation in and control over the process of health care.

ACKNOWLEDGMENTS

The author wishes to thank the Youngstown State University Research Council for providing the funds necessary to conduct this research. Thanks are also due to Dr. Lawrence Weed, Dr. Charles Burger, Lin Turner, patient representatives, caregivers, and to patients at the primary care practice discussed in this paper for their ongoing willingness to respond to my queries about the "knowledge coupler" innovation and the medical practice that employs it.

REFERENCES

Appelbaum, P. S., Lidz, C. W., & Meisel, A. (1987). *Informed consent: Legal theory and clinical practice.* New York: Oxford University Press.

Berg, M. (1997). *Rationalizing medical work: Decision-support technologies and medical practice.* Cambridge, MA: MIT Press.

Berg, M. (1998). The Politics of Technology: On Bringing Social Theory into Technological Design. *Science, Technology & Human Values, 23,* 456–490.

Burger, C. (1997). The Use of Problem Knowledge Couplers in Primary Care Practice. *Healthcare Information Management, 11,* 13–26.

Carter, W. B., Beach, L. R., & Inui, T. S. (1986). The Flu Shot Study: Using Multiattribute Utility Theory to Design a Vaccination Intervention. *Organizational Behavior and Human Decision Processes, 38,* 378–391.

Cegala, D. J., & McGee, D. S. (1996). Components of Patients' and Doctors' Perceptions of Communication Competence During a Primary Care Medical Interview. *Health Communications, 8,* 1–27.

Cegala, D. J. (1997). A Study of Doctors' and Patients' Communication During a Primary Care Consultation: Implications for Communication Training. *Journal of Health Communications, 2,* 169–194.

Charles, C., Gafni, A., & Whelan, T. (1997). Shared Decision-making in the Medical Encounter: What Does It Mean? (Or It Takes at Least Two to Tango). *Social Science & Medicine, 44,* 681–692.

Charles, C., Gafni, A., & Whelan, T. (1999). Decision-making in the Physician-Patient Encounter: Revisiting the Shared Treatment Decision-making Model. *Social Science & Medicine, 49,* 651–661.

Christiansen, C. M., Bohmer, R., & Kenagy, J. (2000). Will Disruptive Innovations Cure Health Care? *Harvard Business Review,* (September–October), 102–112.

Covell D. G., Uman G. C., & Manning P. R. (1985). Information Needs in Office Practice: Are They Being Met? *Annals of Internal Medicine, 103,* 596–599.

Dawes, R. (1988). *Rational choice in an uncertain world.* NY: Harcourt Brace Jovanovich.

Deber, R., Kraetschmer, N., & Irvine, J. (1996). What Role Do Patients Wish to Play in Treatment Decision Making. *Archives of Internal Medicine, 156,* 1414–1420.

Eddy, D. M. (1984). Variations in Physician Practice: The Role of Uncertainty. *Health Affairs, 3,* 74–89.

Eddy, D. M. (1990). Anatomy of a Decision. *Journal of the American Medical Association, 263,* 441–443.

Eisenberg, J. M. (1986). *Doctor's Decisions and the Cost of Medical Care*. Ann Arbor, MI: Health Administration Press.

Elstein, A. S. (1976). Clinical Judgment: Psychological Research & Medical Practice. *Science, 194*, 696–700.

Emanuel, E. J., & Emanuel, L. L. (1992). Four Models of the Physician-Patient Relationship. *Journal of the American Medical Association, 267*, 2221–2226.

Flood, A. B., Wennberg, J. E., Nease, R. F., Fowler, F. J., Ding, J., Hynes, L. M., & the Prostate Patient Outcomes Research Team (1996). The Importance of Patient Preference in the Decision to Screen for Prostate Cancer. *Journal of General Internal Medicine, 11*, 342–349.

Guadagnoli, E., & Ward, P. (1998). Patient Participation in Decision-making. *Social Science & Medicine, 47*, 329–339.

Gustafson, D. H., Hawkins, R., Boberg, E., Pingree, S., Serlin, R. E., Graziano, F., & Chan, C. L. (1999). Impact of Patient-centered, Computer-based Health Information/Support System. *American Journal of Preventive Medicine, 16*, 1–9.

Hall, J., Roter, D. L., & Katz, N. R. (1987). Task Versus Socio-emotional Behaviors in Physicians. *Medical Care, 25*, 399–412.

Haug, M. (1973). Deprofessionalization: An Alternate Hypothesis for the Future. In: P Halmos (Ed.), *Professionalization and Social Change* (pp. 195–211). Keele, Staffordshire: University of Keele.

Hersey, J. C., Matheson, J., & Lohr, K. N. (1997). *Consumer health informatics and patient decision-making*. Rockville, MD: Agency for Health Care Policy and Research (AHCPR), No. 98-N001.

Hunt, R. E., & Newman, R. G. (1997). Medical Knowledge Overload: A Disturbing Trend for Physicians. *Health Care Management Review, 22*, 70–75.

Kaplan, B. (1995). The Computer Prescription: Medical Computing, Public Policy, and Views of History. *Science, Technology, & Human Values, 20*, 5–38.

Leape, L. (1994). Reducing Errors in Medicine. *JAMA, 272*, 1851–1857.

Lesse, S. (1983). A Cybernated Health Diagnostic System – An Urgent Imperative. *American Journal of Psychotherapy, 37*, 451–455.

Lesse, S. (1986). Computer Advances Toward the Realization of a Cybernated Health-Science System. *American Journal of Psychotherapy, 40*, 321–323.

Levine, M. N., Gafni, A., Markham B., & McFarland, D. (1992). A Bedside Decision Instrument to Elicit a Patient's Preference Concerning Adjuvant Chemotherapy for Breast Cancer. *Annals of Internal Medicine, 117*, 53–58.

Leyerle, B. (1984). *Moving and shaking American medicine: The structure of a socioeconomic transformation*. Westport, CT: Greenwood Press.

Maxmen, J. S. (1976). *The post-physician era: Medicine in the twenty-first century*. New York: Wiley.

McKinlay, J., & Arches, J. (1985). Toward the Proletarianization of Physicians. *International Journal of Health Services, 18*, 191–205.

McKinlay, J. B., Potter, D. A., & Feldman, H. A. (1996). Non-medical Influences on Medical Decision-making. *Social Science & Medicine, 42*, 769–776.

Morris, A. H. (2000). Developing and Implementing Computerized Protocols for Standardization of Clinical Decisions. *Annals of Internal Medicine, 132*, 373–383.

Nease, R. F. Jr., & Brooks, W. B. (1995). Patient Desire for Information and Decision Making in Health Care Decisions: The Autonomy Preference Index and the Health Opinion Survey. *Journal of General Internal Medicine, 10*, 593–600.

Ong, L. M. L., de Haes, J. C. J. M., Hoos, A. M., & Lammes, F. B. (1995). Doctor-patient Communication: A Review of the Literature. *Social Science & Medicine, 40*, 903–918.

Parsons, T. (1951). *The social system*. New York: Free Press.

Payer, L. (1988). *Medicine and culture: Varieties of treatment in the United States, England, West Germany, and France*. NY: Penguin Books.

Peersman, W., Jacobs, N., De Maeseneer, J., & Seuntjens, L. (2001). The Flemish Version of a New European Standardised Outcome Instrument for Measuring Patient's Evaluation of the Quality of Care in General Practice. Working Papers in Methodology, Department of Population Studies and Social Science Research Methods, Ghent University.

Reiser, S. J. (1978). *Medicine and the reign of technology*. NY: Cambridge University Press.

Ritzer, G. (2000). *The McDonaldization of society*. Thousand Oaks, CA: Pine Forge Press.

Roberts, C. A., & Aruguete, M. S. (2000). Task and Socioemotional Behaviors of Physicians: A Test of Reciprocity and Social Interaction Theories in Analogue Physician-patient Encounters. *Social Science & Medicine, 50*, 309–315.

Roter, D. L., Hall, J. A., & Katz, N. R. (1987). Relations Between Physicians' Behaviors and Analogue Patients' Satisfaction, Recall, and Impressions. *Medical Care, 25*, 437–442.

Roter, D. L., Stewart, M., Putnam, S. M., Lipkin, M. Jr., Stiles, W., & Inui, T. S. (1997). Communication Patterns of Primary Care Physicians. *JAMA, 222*, 350–356.

Starr, P. (1982). *The social transformation of American medicine*. NY: Basic Books.

Stevenson, F. A., Barry, C. A., Britten, N., Barber, N., & Bradley, C. P. (2000). Doctor-patient Communication About Drugs: The Evidence for Shared Decision Making. *Social Science & Medicine, 50*, 829–840.

Tversky, A., & Kahneman, D. (1974). Judgment Under Uncertainty: Heuristics and Biases. *Science, 185*, 1124–1131.

Voytovich, A. E., Ripey, R. M., & Suffredini, A. (1985). Premature Conclusions in Diagnostic Rasoning. *Journal of Medical Education, 60*, 302–307.

Wason, P., & Johnson-Laird, P. (1972). *Psychology of reasoning*. Cambridge, MA: Harvard University Press.

Weaver, R. R. (1986). Some Implications of the Emergence and Diffusion of Medical Expert Systems. *Qualitative Sociology, 9*, 237–255.

Weaver, R. R. (1991). *Computers and medical knowledge: The diffusion of decision support technology*. Boulder, CO: Westview Press.

Weed, L. L (1991). *Knowledge Coupling: New premises and new tools for medical care and education*. NY: Springer-Verlag.

Weed, L. L. (1994). Reengineering Medicine. *Federation Bulletin, 81*, 149–183.

Weed, L. L. (1995). Reengineering Medicine: Questions and Answers. *Federation Bulletin, 82*, 24–36.

Weed, L. L. (1997). New Connections Between Medical Knowledge and Patient Care. *British Journal of Medicine, 315*, 231–235.

Weed, L. L. (2000). Opening the Black Box of Clinical Judgment – An Overview. *British Journal of Medicine, 319*, 1–4.

Wennberg, J. E. (1984). Dealing with Medical Practice Variation: A Proposal for Action. *Health Affairs, 3*, 6–32.

PART III:
TECHNOLOGY AND PROVIDERS

PREDICTORS OF CONTINUED USE OF TELEMEDICINE BY PRIMARY CARE PROFESSIONALS, MEDICAL SPECIALISTS AND PATIENTS

William Alex McIntosh, John R. Booher,
Letitia T. Alston, Dianne Sykes, Clasina B. Segura,
E. Jay Wheeler, Ted Hartman and William McCaughan

ABSTRACT

Proponents of telemedicine believe this technology will resolve many of the problems associated with the lack of access to specialty care by isolated populations. However, in order for telemedicine to be successful, health care professionals and patients must be willing to use it. Few studies exist that identify those characteristics that differentiate adopters from non-adopters of this technology. Furthermore, little is known about the kinds of health care professionals and patients who are willing to make continued use of telemedicine after initial adoption. Prior studies of the adoption of medical technology have identified personal characteristics such as age and gender among patients and age, gender, years since graduation from medical school, and medical specialty among providers as predictors of the adoption of medical technolgy. Using data collected from the first 483

Changing Consumers and Changing Technology in Health Care and Health Care Delivery,
Volume 19, pages 153–177.

teleconsultations performed by a West Texas medical system, we developed prediction models of the continued use of telemedicine using the personal characteristics of patients, primary care professionals, and medical specialists as predictors. We included the number of primary care professionals and specialists involved in prior teleconsultations in these prediction models. Patient characteristics contributed little to the prediction of continued use of telemedicine; however, characteristics of the health care professionals such as age, gender, and years since graduation as well as the number of health care personnel involved in previous consultations were significant predictors of the continued use of telemedicine.

INTRODUCTION

Telemedicine represents a technology designed to overcome barriers imposed by distance on the transfer of information between patients and medical specialists. At this time, it is primarily used to provide medical personnel practicing in rural areas with needed continuing medical education and access to sources of recent medical research findings. However, its most revolutionary use lies in the application of two-way interactive video for consultations between a patient and a primary care provider located at a remote site and a specialist located at a full-service medical center. In this case the primary care provider seeks clinical information in order to provide better care for patients.

Telemedical technology has been in use since the 1960s, but only recently has its diffusion accelerated (Emery, 1998). Currently, nearly 30% of the nation's rural hospitals have some form of telemedicine and perform, on average, 15 consultations a month (Emery, 1998, 13). Some have argued that until recently the costs of the technology and its implementation have made its adoption unattractive. Others have claimed both organizational requirements and the competitive markets faced by hospitals influence the adoption decision (Anderson & Steinberg, 1994; Emery, 1998). We have argued that while these are important considerations, the adoption and, particularly, the continued use of medical technologies, such as telemedicine, depend upon decisions made by their potential users, i.e. the patients, primary care providers, and specialty care providers. Once the technology is in place, patients and medical staff at both ends of the system must be willing to use it. Telemedicine represents an interesting innovation because its adoption involves, not only providers and their patients, but also consulting specialists in distant locations. Because of the centralized location of most telemedicine equipment, the primary care provider must not only believe a teleconsultation will serve the interests of the patient but also be willing to devote extra time away from other patients in order for such a consultation to take place. Time away from other patients represents potential loss of income as well as an inconvenience for both provider and patient (Institute of Medicine, 1996). A provider may find teleconsultations threatening because the primary care provider might be made to look incompetent in front of her or his patient by the specialist (Institute of Medicine, 1996).

The specialist must also agree to see the patient at a distance instead of in person and face many of the same inconveniences and income impacts as the primary care provider. The specialist's situation is made more complex by the fact, that until very recently, these providers were not reimbursed by Medicaid, Medicare, or traditional private insurance plans unless the consultation involved a face-to-face encounter with the patient (Health Care Financing Administration, 2000). Thus, specialists were essentially asked to donate their time and expertise in such consultations. Some specialists may also be concerned about whether their liability insurance will cover these kinds of consultations (Institute of Medicine, 1996). Adoption and continued use of such technology for consultations, therefore, involves a number of complicating factors.

Finally, the patient must be persuaded that using a teleconsultation is preferable to traveling to see a specialist or to foregoing such contact altogether. The lack of specialists in rural areas sometimes forces rural residents to travel great distances to obtain this level of care. A number of rural residents decline to travel, in part because of resource constraints, and thus forego what is often necessary care from specialists (Ludtke & Ahmad, 1992; McIntosh et al., 2000). Reports of patient resistance to teleconsultations are rare; however, because patient characteristics are associated with the adoption of other medical technologies, such characteristics should not be overlooked in a study of telemedicine.

The factors discussed above act as potential barriers to the initial use of telemedicine technology. However, we contend that once telemedicine technology is used, the experiences of the patients and the providers involved with it will affect decisions regarding its continued use. Studies of the adoption of other types of medical technology provide us with a general idea of the kinds of patients and providers most likely to adopt new technologies. Providers' personal characteristics, past experiences, and salient group memberships all have been shown to influence the adoption decision. However, most adoption research focuses on adoption by individuals and ignores the issue of simultaneous multiple adopters. In the case of medical technology, either patients or providers are studied but rarely are both included in the same study (see Greer, 1981, 1986). We therefore cannot determine with any great certainty whether patient or provider characteristics are more important in these decisions.

The picture becomes even less clear when two sets of care providers are involved in the decision. Finally, most studies of adoption deal only with initial use. The question of individual characteristics and contextual variables associated with continued use are not addressed. This study attempts to utilize patient, primary care provider, and specialist characteristics to predict the continued use of telemedicine in a variety of medical specialties.

LITERATIVE REVIEW

A number of studies of medical technology adoption by providers have been conducted since the 1950s. These studies have found that personal characteristics, such as age, years since graduation from medical school, and specialty, affect adoption (Fendrick & Schwartz, 1994). A number of social characteristics also impact adoption. These characteristics include the nature of the doctor's practice (e.g. solo versus group), the nature of the doctor's professional network, and the centrality of the doctor in such networks (Burt, 1987; Strang & Tuma, 1993). In general, the older the provider, the less he or she is willing to take the risks involved in adopting new medical technology. Rogers (1962) speculated in an earlier version of his classic work on adoption-diffusion that age should be considered a cohort effect. People growing up in a particular cohort become familiar and comfortable with the technology of the period. Medical school exposes students to the latest advancements in medical technology. As a result, providers graduate and practice medicine with a degree of expertise and comfort in using this medical technology. As technology continues to advance, it becomes less like the technology they learned about and became comfortable with when they were young (see Greer, 1986 for a similar conclusion). However, not all such studies have found an inverse relationship between age and adoption. Some have found that older providers adopt new medical technology earlier than their younger colleagues (Becker, 1970). In a study of doctors' professional networks and medical adoption, Anderson and Jay (1985) found that the group most likely to adopt and continue to use computer assisted medicine were older and were more actively engaged in professional activities outside their hospital. They were also more likely to be medical generalists rather than specialists and sources of information for other doctors in the hospital. Studies of agricultural adoption may provide insights into the characteristics of patient adopters. In this literature, young farmers are thought to be greater risk takers. Regarding preventive health care practices, males generally take greater risks with their health than females (Ross & Bird, 1994). Thus, females may lead males in the use of telemedicine, because telemedicine represents an improved opportunity to access medical care. Yet at least one study found that males are more likely to choose to use telemedicine for particular medical conditions than are females (McIntosh et al., 2000).

Early research indicated that minorities had higher levels of mistrust of the medical profession; recent studies have found, however, that minorities use health care services at the same rate as the majority (Aday & Awe, 1997). McIntosh et al. (2000) also found no ethnic differences in patients' self-reported willingness to use telemedicine under hypothetical conditions. Patient adopters may also have characteristics similar to those of persons most likely to utilize health care.

Thus, research based on the health behavior model becomes relevant. This model suggests that the use of health care services depends upon the balance between a set of facilitators and barriers to use. These are conceived of in more specific categories including predisposing, enabling, and need factors. Predisposition captures those social, cultural, and psychological propensities to seek formal care. Gender, ethnicity, education, and region are the pertinent barriers/facilitators to the use of health care by individuals. Enabling factors are those resources, or lack thereof, that make it possible to take advantage of existing health care. A related factor here is location; distance to health care may provide an additional barrier for some, particularly when convenient sources of transportation do not exist. Finally, the need for health care may be represented by an existing condition or by a subjective judgement regarding health. Need tends to have the greatest influence on seeking health care.

METHODOLOGY

Data

The present study uses data from a large number of teleconsultations, involving a variety of patients, primary care providers, and specialists. All of these individuals in effect adopted the technology; some adopted later than others; some used the technology only once; others used it multiple times.

Data for this study were compiled from records of all consultations conducted by two-way interactive video between a rural hospital, with a small number of primary health care providers, and an urban university medical center between September 1, 1990 to June 6, 1996. During this 6.5-year-period, 483 consultations took place. Even though records were maintained by both the rural hospital and the urban medical center, not all consult records were complete. Nevertheless, virtually complete information was available on the following

Table 1. Descriptive Characteristics of Primary Care Providers Using
Telemedicine.

Primary Care Provider Characteristics

Professional Area	Number of Providers in Professional Area	Number of Consultations in Professional Area	Percent of Total
Family Practice	8	457	95.4
General Practice	5	10	2.1
Nursing (RN)	4	10	2.1
Obstetrics	1	1	0.2
Pediatrics	1	1	0.2
Total	19	479*	100.0

Age
Mean	37.9
Standard Deviation	5.1

Gender
Number Male	18
Percent Male	95.4

* These data were not available for all consultations.

variables: patient age, gender, county of origin, and complaint; primary care
provider age, gender, and type of medical specialty (e.g. family practitioner,
registered nurse, provider's assistant, etc.); and specialist age, gender, specialty
type, and year of graduation from medical school.

Description of Population
At the rural hospital end, 95% of the consultations were performed by family
practitioners, while 4% of the consultations were performed by general practitioners and registered nurses (see Table 1). The average age of the consultant
was 37.9 years, and 95.4% were male. Patients averaged 40.4 years in age, and
45.5% were female (see Table 2). Had they been charged, 58.4% of the patients
reported that they would have paid for the consultations from their own pockets.
More than 86% of these patients participated in a single teleconsultation.
Slightly more than 1% participated in four or more consultations. The average
age of the urban specialists was 44.5 years, and on average they had left medical
school nearly 14 years earlier (see Table 3). Eighty percent of the specialists

Table 2. Descriptive Characteristics of the Patients Using Telemedicine.

Patient Characteristics

Age
Mean	40.4
Standard Deviation	28.8

Gender
Number Male	254
Percent Male	54.5

Consultations Per Patient	Frequency	Percent
One	417	86.3
Two	45	9.3
Three	14	2.9
Four	5	1.0
Five	2	0.4
Total	483	100.0

were male and 33.2% of the consultations were performed by dermatologists and 10% were performed by orthopedists.

The urban specialists and the rural primary care providers did not make equal use of the telemedicine system. Some made much more frequent use than others. That is, one of the dermatologists participated in nearly 25% of the consultations, and two of the rural primary health care providers were separately involved in nearly 80% of the consultations.

Analysis Technique

We were interested in predicting the continued use of telemedicine by primary care professionals, medical specialists, and patients over time. Not all of those who used the telemedicine system began using it at the same point in time. Some of the medical personnel began to use the equipment as soon as it was on line; others made use of it well past the time it was first available. Some of these late users delayed use because they either lacked the need for telemedicine or because they joined the medical system in West Texas at a later time period. The same can be said of course of the patients; they made no use of telemedicine either because of lack of need or late entry into the West Texas

Table 3. Descriptive Characteristics of the Medical Specialists Using
Telemedicine.

Specialist Characteristics

Specialty	Number of Specialists	Number of Consultations	Percent
Dermatology	7	145	33.2
Surgery	8	45	10.3
Pediatrics	17	45	10.3
Orthopedics	9	42	9.6
Endocrinology	8	33	7.6
Neonatology	7	21	4.8
Other*	27	105	24.1
Total	93	436**	100.0

Age
Mean	44.5
Standard Deviation	12.0

Gender
Number Female	20
Percent Female	21.2

* Includes allergy, cardiology, nephrology, urology, oncology, genetics, gynecology, obstetrics, neurology, burns, radiology, rheumatology, pulminology, gastroenterology. Each of these specialties represents 2% or fewer of the consultations
** These data were not available for all consultations.

medical system. In technical terms, our time series data were "censored" because of the lack of uniformity in exposure to the technology, thus making ordinary least squares regression models inappropriate for our analysis (Allison, 1984). An appropriate alternative is "event history analysis" via Cox regression. This type of analysis is used to study mortality in populations over time, where mortality acts as the censoring agent. "Event history data provide information on the times during a specified interval (i.e. a continuous observation period) when members of a sample change from one discrete state to another, plus the sequence of states they occupy" (Mayer & Tuma, 1990, 9).

In order to compare primary health care providers on the duration of time elapsed between consultations, a fixed effects version of Cox regression was used because participation in a teleconsultation was an event that could be repeated. Cox regression is the multivariate technique of choice in a time series in which the time interval between events varies. A fixed effects model has both advantages and disadvantages. An advantage of the fixed effects model is all variables that are constant over time are, in effect, controlled for in the modeling. The disadvantage of the fixed effects model are no variables that are constant across time for a group of interest can be included as independent variables. In the case of the present analysis, we first consider the change in probability that a primary care provider will participate in another consultation. Thus, variables characterizing primary care providers that are constant over time (such as gender) cannot be used in the analysis. However, the gender of the patients or of the specialists can be used in this particular analysis.

For our analysis, the dependent variable in Cox regression was the amount of time from the availability of the teleconsultation equipment until the actual use of that equipment; the time duration was measured in days. Even though the Cox regression model has the actual number of days as the dependent variable, it orders events according to when they happened (Allison, 1995; DeLong et al., 1994; Farewell & Prentice, 1980). In order for this method to work well, ties (events that occur on the same day) must be relatively rare. Fortunately, ties were rare in this data set. However, when more than one event occurred on a single day, an estimate was made regarding which one probably came first. The dependent variable structure is further specified by the use of a censoring agent. In this case, the censoring agent is the use of the teleconsultation equipment to obtain information from a particular specialty area, such as surgery.

For example, in the case of surgery, the Cox regression model is comparing the time duration of the use of the teleconsultation equipment to obtain information about a potential surgical case to the time duration of the use of the same equipment to obtain information about another, non-surgical type of case. It accomplishes this task by calculating the probability in which an event (an event is the use of the teleconsultation equipment) might require surgical consultation based on the distribution of events that might call upon another specialty including surgery. These probability calculations are based on the order and number of events in each category (surgery vs. all other specializations). The time duration of the events is detailed so the events will be in their proper order for calculating the probability that an event might involve a particular specialty, such as surgery. By multiplying that probability by the total number of units of time that characterize the observation period, the amount of change in the chances that a particular specialty will be involved in consultations is obtained.

Because of the limitations imposed by the fixed effects model, we ran a second set of analyses, stratifying on specialty areas instead of primary care provider. Such models allowed us to include primary care provider variables that are constants; thus gender of the primary care provider could be explicitly included in the modeling effort.

RESULTS

Table 4 contains the results of the fixed effects Cox regressions for each area of specialty consultation drawn upon by primary care providers (note that specialist characteristics are fixed in these models), and Table 5 contains similar information for specialists (primary care provider characteristics are fixed in these models). Each table includes the appropriate risk ratios, R squares, and the Wald chi-squared statistics. A Wald chi-squared is used because of its conservative nature in that Type I error is less likely than with alternative techniques. The R squares are not true measures of variance explained; rather they measure the impact of the set of independent variables in the equation. Risk ratios are similar to odds ratios and discussed below. In each equation, the dependent variable is the comparison of time duration of using the tele-consultation equipment for a given specialty versus using that equipment for any other specialty.

Risk ratios are positive if they are greater than one and are negative if they are less than one. The easiest way to interpret a given risk ratio is to subtract 1.0 and then multiply the result by 100 to form a percentage. For example, the risk ratio associated with the impact of primary care providers' age on neonatology is 0.932. Subtracting 1.0 from that figure gives a negative 0.068. Multiplying this number times 100 gives −6.8%. Thus, for each year of age of a primary care provider, the probability of another consultation involving neonatology decreases by 6.8%.

The chi-squared for each equation is moderately high as are the Rs, which range from 0.35 for dermatology to 0.47 for both pediatrics and other specialties. Each equation has a chi-squared that has an associated alpha of 0.01 or less. Consequently, goodness of fit is acceptable. Figures 1–4 illustrate the findings from the Cox regressions for the primary care providers; Figures 5–8 provide similar information for the specialists. Table 4 and Fig. 1 show that for each year of age of the primary care provider, the likelihood of another consultation declined. In the case of neonatology, the decline is on the order of 6.8%, for pediatric patients, the decline is 5.7%. The cumulative effects of these declines over the seven year observation period is substantial; the decline in probability for pediatrics, for example, was 39.9% (7 × 5.7%).

Figure 3 compares the impact of the number of specialists involved in a consultation on the probability of another consultation taking place. Consultations

Table 4. Fixed Effects Model of Cox Regressions: Stratified by Specialist.

Variable	Neonatology	Pediatrics	Dermatology	Surgery	Orthopedics	Endocrinology	Other Specialties
Primary Care Provider							
Age	0.932**	0.943**	0.909***	0.920**	0.918**	0.914***	0.925***
Patient Age	0.997	0.997	1.008	0.996	0.997	.0995	0.998
Patient is Male	0.915	0.800	1.431	0.826	0.821	0.784	0.879
Years Since Graduation:							
Specialist	0.998	0.989	0.957	0.993	0.995	0.996	0.995
Number of Primary Care Providers	3.089***	3.123**	2.707	3.104**	3.147**	3.343**	3.256**
Number of Specialists	11.501***	8.783***	9.206***	8.346***	9.349***	5.770***	8.087***
Number of Consults by Primary Provider	0.992***	0.991***	0.988***	0.992***	0.992***	0.991***	0.991***
R Squared	0.41	0.47	0.35	0.42	0.42	0.44	0.47
Chi Square (7 degrees of freedom)	88.000***	99.362***	54.766***	86.453***	89.032***	88.796***	100.173****

* Alpha < 0.05, ** Alpha < 0.01, *** Alpha = 0.0001

Censoring Agent (on) 1 = Specialty was required during teleconference
Censoring Agent (off) 0 = Specialty was not required during teleconference
Figures in table are risk ratio unless otherwise specified.

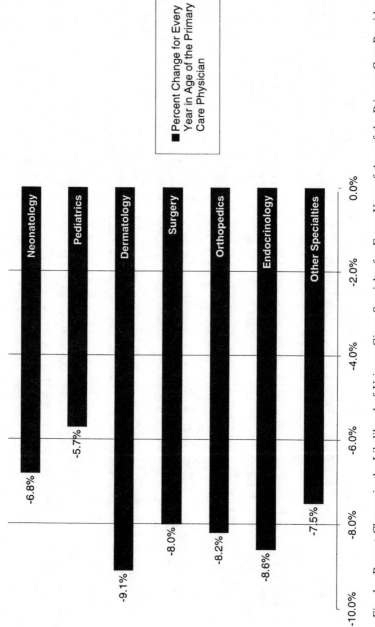

Fig. 1. Percent Change in the Likelihood of Using a Given Specialty for Every Year of Age of the Primary Care Provider: Stratified by Specialist.

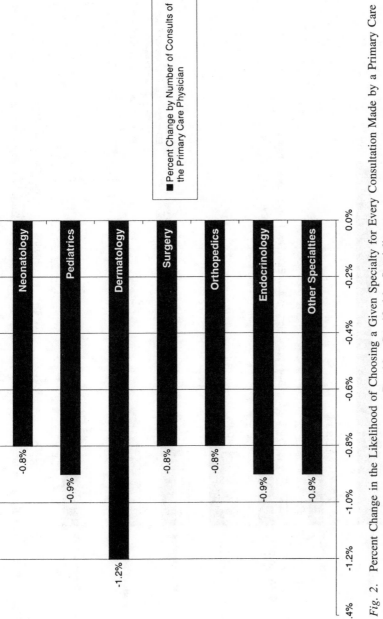

Fig. 2. Percent Change in the Likelihood of Choosing a Given Specialty for Every Consultation Made by a Primary Care Provider: Stratified by Specialist.

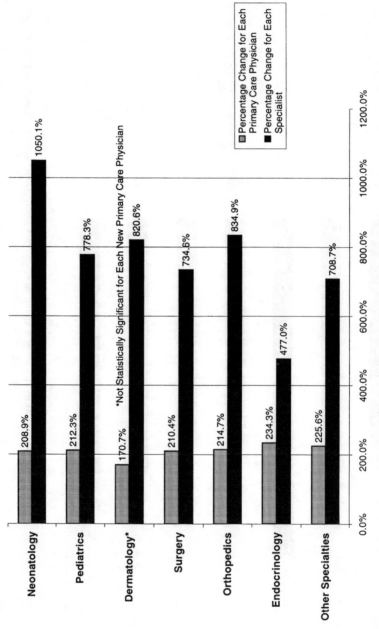

Fig. 3. Percentage Change in the Likelihood of Selecting a Given Specialty for Each New Primary Care Provider Involved in the Diagnosis: Stratified by Specialist.

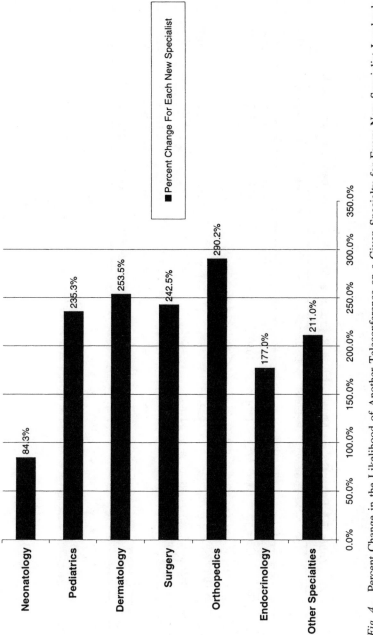

Fig. 4. Percent Change in the Likelihood of Another Teleconference on a Given Specialty for Every New Specialist Involved: Stratified by Primary Care Provider.

Table 5. Fixed Effects Model of Cox Regressions Stratified by Primary Care Providers.

Variable	Neonatology	Pediatrics	Dermatology	Surgery	Orthopedics	Endocrinology	Other Specialties
Primary Care Provider Age	0.358***	0.367***	0.337***	0.362***	0.375***	0.357***	0.350***
Patient Age	1.000	1.001	0.997	0.996	0.996*	0.997	0.997
Years Since Graduation:							
Specialist	1.002	1.005	1.014*	0.994	0.984*	1.010	1.004
Number of Primary Care Providers	2.200*	2.071*	1.239	1.883	1.663	2.209*	2.054*
Number of Specialists	1.182	1.509*	1.207	1.452	1.891**	1.358	1.385
Number of Consults by Primary Provider	1.000	1.000	0.964***	1.002	1.004*	0.999	0.999
R Squared	0.86	0.86	0.88	0.86	0.84	0.86	0.90
Chi Square (7 degrees of freedom)	243.607***	244.809***	217.670***	255.428***	236.680***	240.342***	287.287****

* Alpha < 0.05, ** Alpha < 0.01, *** Alpha = 0.0001

Censoring Agent (on) 1 = Specialty was required during teleconference
Censoring Agent (off) 0 = Specialty was not required during teleconference.

involving multiple specialists were infrequent (fewer than 10% included more than one), but their impact on the probability of further consultations was quite high. For example, the probability of another neonatal consultation increased 1,050% if more than one specialist was involved in a given consultation. Figure 3 also contains the percentage change in probability if multiple primary care providers were involved in the consultation. Again the effect is positive, although not as substantial as the effect for multiple specialists. If more than one primary care provider was involved in a neonatal consultation, the probability increased by 209% that another neonatal consultation would take place. However, as in the case of multiple specialists, the percentage of consultations that involved more than one primary care provider was low.

As mentioned above, we also stratified the Cox regression equations by primary care providers (see Table 5 and Figs 4–8). Figure 4 again demonstrates the impact of multiple specialists on the probability of more consultations in a given specialty area.

The gender of the specialist made a difference in the probability of additional consultations as well. In the case of neonatology, the probability of another consultation increased 104% when the specialist was male; for endocrinology, the increased probability was 243%. When the gender of the patient was male, the probability of another consultation declined from 19% for pediatric cases and 31% for those involving dermatology. Years since graduation of the specialists had an effect similar to that of primary care provider age: it led to a decline in the probability of another consultation in a given specialty. The declines were modest and not statistically significant in the cases of dermatology and endocrinology. For example, the probability of another consultation involving orthopedics declined 2.8% for every additional year since graduation on the part of orthopedic specialists. During the nearly seven-year observation period, this decline amounted to 19.6%.

CONCLUSIONS

Some of our findings substantiate those of others regarding medical professionals' adoption of medical technology. This is particularly true of age and years since graduation. Some researchers have found that among providers in private practice, younger doctors and doctors who have more recently graduated from medical school adopt innovations more rapidly than older, less-recently graduated doctors. Other investigators, however, have observed a greater propensity for older providers to innovate more quickly than younger ones when the context is a group practice of some sort. The tendency of younger, more

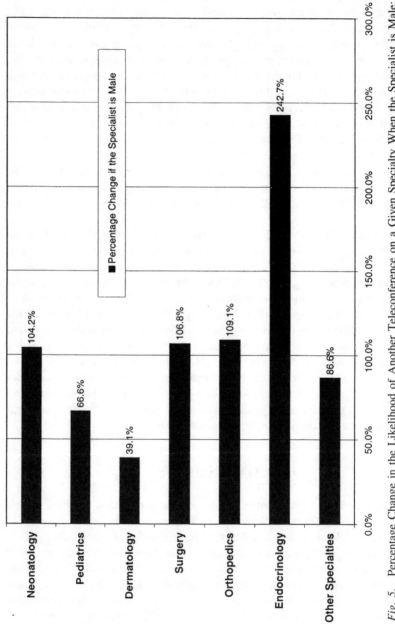

Fig. 5. Percentage Change in the Likelihood of Another Teleconference on a Given Specialty When the Specialist is Male: Stratified by Primary Care Provider.

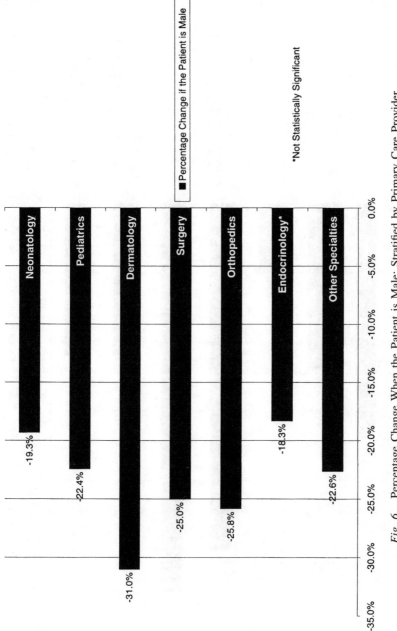

Fig. 6. Percentage Change When the Patient is Male: Stratified by Primary Care Provider.

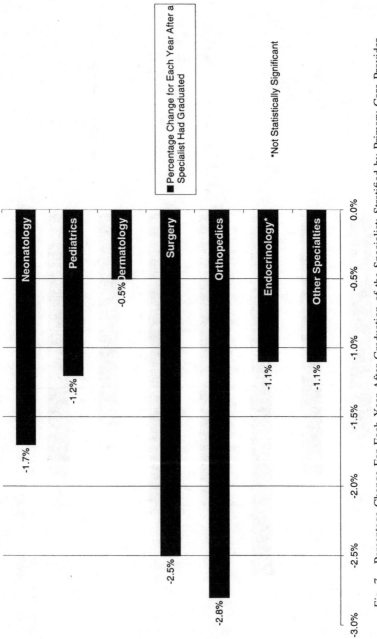

Fig. 7. Percentage Change For Each Year After Graduation of the Specialist: Stratified by Primary Care Provider.

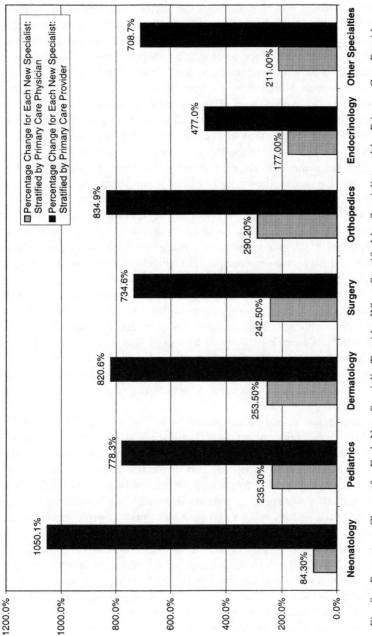

Fig. 8. Percentage Change for Each New Specialist/Provider When Stratified by Specialist and by Primary Care Provider.

recently graduated providers to adopt the telemedicine technology earlier than their older colleagues is consistent with our assessment of situational pressures at both the specialty and primary care ends of the telecommunications link. The specialists in our study were located at a university medical center and were encouraged to do teleconsultations in this setting. Similarly, the primary care professionals were affiliated with a hospital whose survival, in part, depended on increasing the patient base of the hospital (McIntosh et al., 1999). Older primary care providers were less likely to use teleconsultations over time, and the longer it had been since a specialist graduated from medical school, the less likely she or he was to participate in any teleconsultation.

The greater willingness among younger primary care providers to participate in teleconsultations as shown by our data is consistent with normal provider practice. Younger primary providers are less experienced than older providers in dealing with a wide spectrum of patient ills and are also less likely to have developed relationships with colleagues in the various specialty areas. Participation in the teleconsultation system provided additional training by exposing primary care professionals to ailment variety and the process of specialists' diagnoses. Participation could also be seen as an advantage since these young providers were not forced to choose between specialists available for consultation at the distant teaching institution and an established network of specialists with whom the primary care provider was used to dealing. The young providers were less likely to be locked into a set of previously established consultative relationships. Thus, the telemedicine link provided young primary care providers with a means to acquire further diagnostic and treatment skills and also access to specialists to whom they could refer future patients when necessary.

As among the primary care providers in our sample, the probability of specialists participating in another consultation also depended on how long ago they had left medical school; less recently graduated providers showed more rapid and more frequent use. Part of the reason for their availability for teleconsultations may have to do with the time it takes to establish a practice and a research program. It is possible that older, well-established specialists were less available for consultations because of prior commitments to their own patients or to their research. Shaperman and Backer (1995) found that academic medical centers often resisted new knowledge because of commitments to traditional practice methods and a lack of incentives to do so, among other things. Those faculty employed longer in such settings are more likely to be affected by such a context. However, in this particular case, those who determined the specialty or specialties needed for a particular consult and who scheduled these consultations with the specialists have stated that they determined which individual specialists they believed to be best qualified as a consultant and sought that person out. As years since graduation decreased the likelihood of further consultations, this suggests that the older specialists were thought to be more qualified and were sought first during the initial stages of the system's implementation. Less experienced specialists were brought into the consultations after the system was well established.

Of equal interest is the lack of patient effects on telemedicine's use. Neither age nor gender mattered when it came to the continued use of telemedicine. This finding suggests that teleconsultations were made available to and were used by a variety of patients, We believe, however, that other patient characteristics may have influenced their participation, the most important of which is medical condition. Clearly, some of the patients made additional uses of telemedicine because their conditions warranted further consultations. Unfortunately, the patient registry from which we drew our data contained information on the initial complaint/diagnosis for less than half of the patients. It would also have been useful to have had data on patient characteristics such as education and income; unfortunately, the patient registry lacked this information.

An additional finding was teleconsultations that involved multiple specialists were more likely to lead to further consults. This may indicate several things. First, a medical complaint may appear serious but insufficiently clear for diagnosis, requiring a number of different specialists to reach a diagnosis and treatment plan. These more difficult diagnostic situations may require further consultations. A second possibility is that the consultation exposes more specialists to the telemedicine system, increasing their likelihood of using it again. Interviews with the specialists and others indicate that few specialists demonstrated reluctance to use the telemedicine system, so this second interpretation seems less likely.

Another finding of interest is the differential probability of further use by the specialties. Part of this may have been the result of the kinds of cases the system encountered during this time period and part may be due to the propensity of some specialties to make greater use of follow-ups. Dermatology, for example, tends to require more follow-up consultations than do a number of other specialties.

The decline in telemedical consultations that occurred during the study period does not necessarily reflect dissatisfaction with the system. In fact, most of the providers we interviewed indicated a high level of satisfaction with the system. Some did indicate that their satisfaction declined to some degree because of administrative changes in the system over time. The most likely explanation of the decline in consultations was associated with its educational functions, particularly for the primary care professionals. In fact, the two providers who made the most frequent use of telemedicine in each of the years we studied indicated they used the system less because they needed it less. They had learned to diagnose many of the conditions for which they initially had required consultations. If teleconsultation has positive consequences for the medical continuing education of primary care providers in remote locations, this contribution should be recognized. Furthermore, criteria used to ascertain the value and impact of teleconsultation should include this aspect of its effect.

Finally, telemedicine may spread to other remote areas, not only because of the success of the current system, but also because of the growing rural manpower shortage in a number of medical fields (Center for Rural Health Initiatives, 1995).

ACKNOWLEDGMENTS

This study was supported by grant number R01 HS08247 from the Agency for Health Care Policy and Research (AHCPR).

REFERENCES

Aday, L. A., & Awe, W. C. (1997). Health service utilization models. In: D. S. Gochman (Ed.), *Handbook of Health Behavior Research, Vol. 1. Personal and Social Determinants* (pp. 153–172). New York: Plenum.

Allison, P. D. (1984). *Event history analysis: regression for longitudinal event data.* Newbury Park: Sage.

Allison, P. D. (1995). *Survival analysis using the SAS system: a practical guide.* Cary: SAS Institute.

Anderson, G. F., & Steinberg, E. P. (1994). The role of the hospital in the acquisition of technology. In: A. C. Gelijins & H. V. Dawkins (Eds), *Medical Innovation at the Crossroads, Vol. IV. Adopting New Medical Technology* (pp. 61–70). Washington, D.C.: National Academy Press.

Anderson, J. G., & Jay, S. J. (1985). The diffusion of medical technology: social network analysis and policy research. *Sociological Quarterly, 26,* 49–64.

Becker, M. H. (1970). Sociometric location and innovativeness: reformualtion and extension of the diffusion model. *American Sociological Review, 34,* 267–282.

Burt, R. S. (1987). Social contagion and innovation: cohesion versus structural equivalence. *American Journal of Sociology, 92,* 1287–1335.

Center for Rural Health Initiatives (1995). *Rural health in Texas: A report to the Governor and the 74th Texas Legislature.* Austin, TX.

DeLong, D. M., Guirguis, G. H., & So, Y. C. (1994). Efficient computation of subset selection probabilities with application to Cox regression. *Biometrica, 81,* 607–611.

Emery, S. (1998). *Telemedicine in hospitals: issues in implementation.* New York: Garland Press.

Fendrick, A. M., & Schwartz, J. S. (1994). Physicians' decisions regarding the acquisition of technology. In: A. C. Gelijins & H. V. Dawkins. (Eds), *Medical Innovation at the Crossroads, Vol. IV Adopting New Technology* (pp. 71–84). Washington, D.C.: National Academy Press.

Farewell, V. T., & Prentice, R. L. (1980). The approximation of partial maximum likelihood with emphasis on case control studies. *Biometrica, 67*, 273–278.

Goldsmith, J. (2000). How will the internet change our health system? *Health Affairs, 19*, 148–156.

Greer, A. L. (1981). Medical technology: assessment, adoption, and utilization. *Journal of Medical Systems, 5*, 129–145.

Greer, A. L. (1986). Medical conservatism and technological acquisitiveness: the paradox of hospital technology adoption. *Research in the Sociology of Health Care, 4*, 185–235.

Healthcare Financing Administration (2000). Program Memorandum – Carriers B-00-02. HCFA-Pub. 60B. Washington, D.C.: Department of Health and Human Services.

Institute of Medicine (1996). *Telemedicine: a guide to assessing telecommunications in health care.* Washington, D.C.: National Academy Press.

Ludtke, R. L., & Ahmad, K. (1994). Sparse populations and patterns of health care use. *Texas Journal of Rural Health*, (first quarter), 8–17.

Mayer, K. U., & Tuma, N. B. (1990). Life course research and event history analysis: an overview. In: K. U. Mayer & N. B. Tuma (Eds), *Event History Analysis in Life Course Research* (pp. 3–29). Madison: University of Wisconsin Press.

McIntosh, W. A., Booher, J. R., Alston, L. T., Sykes, D., & Segura, C. B. (2000). Predictors of use of telemedicine for differing medical conditions. In: J. J. Kronenfeld (Ed.), *The Sociology of Health Care* (Vol. 17, pp. 199–213). Stamford: JAI Press.

McIntosh, W. A., Alston, L. T., Sykes, D., & Segura, C. B. (1999). Rural hospital survival: a case study of the shaping and effect of community perceptions. In: J. J. Kronenfeld (Ed.), *The Sociology of Health Care* (Vol. 16, pp. 223–241). Stamford: JAI Press.

Rogers, E. (1962). Diffusion of innovations. New York: Free Press.

Ross, C. E., & Bird, C. E. (1994). Sex stratification and health lifestyle: consequences for men's and women's perceived health. *Journal of Health and Social Behavior, 35*, 161–178.

Shaperman, J., & Backer, T. A. (1995). The role of knowledge utilization in adopting innovations from academic medical centers. *Hospital and Health Services Administration, 40*, 401–413.

Strang, D., & Tuma, N. B. (1993). Spatial and temporal heterogeneity in diffusion. *American Journal of Sociology, 99*, 614–639.

MANAGING MENTAL ILLNESS: TRENDS IN CONTINUING MENTAL HEALTH EDUCATION FOR FAMILY DOCTORS, 1977–1996

Terri A. Winnick, Regina E. Werum and
Eliza K. Pavalko

ABSTRACT

Medical sociologists have documented psychiatry's tendency to define psychological and behavioral problems as mental disorders. But we know little about how other medical specialties view mental disorders and their treatment. We explore trends in mental health CME (continuing medical education) by analyzing twenty years of data on the quantity and content of articles on mental illness in family practice journals (1977–1996). Pooled time series analyses indicate that drug innovations had the strongest influence on the number of mental illness articles published. New drugs sharply increased attention to all types of mental disorders, not just those with pharmaceutical treatment indications. DSM modifications influenced publications in a complex way: The expanding number of medicalized illnesses in the DSM had a positive impact, while the paradigm shift to the medical model with DSM-III actually decreased the number of articles. Similarly, the supply of patients generated by deinstitutionalization increased cover-

Changing Consumers and Changing Technology in Health Care and Health Care Delivery,
Volume 19, pages 179–203.
ISBN: 0-7623-0808-7

age of mental illness in this literature, while the growth of managed care was associated with a decrease therein. Content analysis of articles on serious mental illnesses (n = 202) shows that drug innovations only increased the number of articles but did not affect drug treatment recommendations. These trends reflect how the medical model and drug treatments have become firmly established in the management of mental illness.

INTRODUCTION

Structural changes in the financing and treatment of mental illness over the past 25 years have dramatically altered the landscape of mental health care in the United States. Social scientists have carefully documented major shifts such as change in the location of treatment (Brown, 1985; Grob, 1991; Mechanic, 1989; Wegner, 1990) and change in the philosophy and scope of diagnosis (Kirk & Kutchins, 1992; Kutchins & Kirk, 1997; Rogler, 1997). But much less attention has been paid to change in the provision of care. Non-psychiatric professionals, most notably family practitioners, now routinely treat about half of all persons diagnosed with a mental health problem (Orleans et al., 1985; Schurman, Kramer & Mitchell, 1985). Their involvement in mental health care extends far beyond acting as gatekeepers who refer cases to psychiatrists or other mental health professionals (Falloon & Fadden, 1993; Goldberg & Huxley, 1980; Mechanic, 1989; Miranda et al., 1994; Sartorioius et al., 1990). Indeed, mental health researchers often refer to them as the "de facto mental health system" (Regier, Goldberg & Taube, 1978).

The critical role of primary care physicians in treating mental illness raises questions about the priorities of practitioners who provide much of mental health care (Cook & Wright, 1995), how they view mental illness and its treatment, and how this perspective may have changed over time. What kind of mental health education do general practitioners receive? Does the amount and type of information change over time? If so, do these trends correspond to paradigmatic changes in psychiatric treatment, regarding diagnostic and fiscal priorities or pharmaceutical innovations?

We address these questions by examining one type of continuing medical education (CME) provided to physicians. Medical doctors are expected and encouraged to engage in CME to enable them to keep abreast of new developments in their field. This education is provided through medical conferences, seminars, and in articles published in the professional literature. We focus on medical journal articles and draw data from articles published in two major family practice journals between 1977 and 1996. We analyze trends over time in the quantity and content of articles on mental disorders in these journals. Family practice was examined because it is the specialty established expressly to provide broad comprehensive medical care to the entire family by "integrating the biological, clinical and behavioral sciences" (American Board of Medical Specialties, 1999, 907). It is also a relatively new specialization, only formally recognized by the National Academy of Medical Specialists in 1975. We track attention to mental disorders during a time period that represents substantial philosophical and treatment shifts in psychiatry, transformations in the financing of medical care, and the emergence of the family practice specialty.

Our analyses seek to answer three questions. *First, how much information do these journals provide about mental illness and has the amount of information changed over time?* To answer this question, we count the number of articles on mental disorders, categorize them by type of disorder, and display these trends over the twenty-year period. We posit that increases in the number of articles represent increasing interest in mental illness; a decrease reflects either a decline in general interest in mental illness or a degree of routinization in mental health treatment

Second, to what extent have structural changes in medical care and psychiatry influenced the number of articles published on mental illness? We are particularly interested in four major structural changes: deinstitutionalization, the growth of managed care, revisions in the *DSM*, and drug innovations, because these have altered priorities regarding the location, financing, and treatment of mental disorders. We present two sets of pooled time series analyses of these four processes: (a) on the total number of mental health articles, and (b) on the number of articles commonly considered the three most severe and persistent mental illnesses, namely schizophrenia, major depression and bipolar disorder. We singled out these serious mental illnesses (hereafter SMI) for two reasons. SMI disorders are the most disabling and the most likely to be regarded as a problem relevant to medical practitioners, especially because specific medical treatment indications exist. Moreover, treatment of SMIs is more likely than treatment of other disorders to be affected by the above-mentioned structural trends.

Third, has the content of mental health articles also shifted over time? We are particularly interested in exploring whether drug innovations and the shift to the medical model marked by the DSM-III alter the content of the articles.

Empirical Expectations

A number of important health care changes have altered where patients receive mental health care, who provides that care, and the nature of the treatment provided. Four trends are particularly germane to family practitioners and thus likely to influence the amount of journal attention to mental health. *Deinstitutionalization* and the growth of *managed care* alter the supply of patients with mental disorders. *Change in the DSM* marks the theoretical shift within psychiatry and broadens the scope of problems defined as medically treatable (Brown, 1995; Conrad, 1992; Kirk & Kutchins, 1992; Kutchins & Kirk, 1997). *Drug innovations* represent a change in treatment technology, providing family practitioners with a set of treatments that are viewed as more familiar, accessible, and effective than other options previously available (Healy, 1997). In this section we briefly review each of these changes and suggest how they may influence trends in the amount of information in family practice journals.

Deinstitutionalization

Prior to the1960s, psychiatric treatment in the United States took place primarily in institutional settings, with as many as a half million persons in long-term custodial care. Since then, the location of treatment for most patients has been shifted to the community. Deinstitutionalization was a gradual process, occurring over a period of years. Declines in patient counts were greatest in the late 1960s and early 1970s but continued at a reduced rate into the 1990s. In 1977, when our data begin, state hospitals still housed 160,000 patients, but by 1994, only 70,000 inpatients remained (U.S. Bureau of the Census, 1980, 1998). Community treatment in a wide range of settings has for the most part replaced custodial care, and now many patients rely entirely on general practitioners to provide treatment (Goldberg & Huxley, 1980; Mechanic, 1989; Miranda et al., 1994). Because deinstitutionalization increases the potential supply of persons with severe mental illness to family practitioners, we expect the amount of attention to mental illness, especially SMI (schizophrenia, major depression, bipolar disorder), to increase accordingly in family practice journals.

Managed Care

The growth of managed care presents the second factor influencing primary care physicians' supply of patients with mental disorders. Contractual care first developed in the late 19th century but did not gain momentum until the 1960s and early 1970s when President Nixon signed into law the Health Maintenance Act of 1973 to promote health maintenance organizations (HMOs) (Mayer & Mayer, 1985; Starr, 1982). In 1996, more than 59 million people, or almost one fourth of the U.S. population, were enrolled in 630 HMOs (National Center for Health Statistics, 1998). This presents nearly a tenfold increase since 1977 (U.S. Bureau of the Census, 1987).

The explosion in managed care may affect the supply of patients to family doctors in several ways. In general, reliance on family practice for most medical treatment has increased because family practitioners provide much of the primary care encouraged by managed care programs. But the expansion of managed care for mental health treatment has been slower and more uneven than for regular medical care. One of the legacies of the psychoanalytic era was the exclusion of insurance coverage from psychiatric treatment, which was seen as a costly and time-consuming luxury (Bittker, 1992). Following this tradition, many managed care programs restrict care for chronic mental illness, or if provided, often refer mental health treatment to special mental health "carve out" programs (Bennett, 1988; Bittker, 1985; McFarland, 1994). While we expect increases in managed care enrollments to have an impact on the number of articles focusing on mental disorders, the direction of the effect is unclear. Managed care simultaneously increases the supply of patients to family doctors for general medical care while at the same time directing mental health treatment in the most severe and prolonged cases to other providers.

Revisions in the DSM
Since the 1970s, American psychiatry has substantially redefined mental disorders, moving away from psychodynamic towards biomedical explanations (Kirk & Kutchins, 1992; Rogler, 1997). This process of medicalization has been formalized through revisions of the *Diagnostic and Statistical Manual of Psychiatric Disorders* (American Psychiatric Association, 1994). The *DSM* was first published in 1952, with new editions issued in 1968, 1980, 1987 and 1994. The latest edition, the *DSM-IV*, identifies and describes more than 350 diagnosable conditions, an increase in coded categories of more than 64% between 1977 and 1996.

These efforts to medicalize mental illness lead us to expect a complex pattern regarding the amount of attention family practice journals pay to mental health issues. On one hand, we expect that revisions in the DSM will produce a short-term increase in the amount of attention paid to mental illness. After all, as each new version of the DSM expands the range of problems defined as disorders, the number of problems defined as medically treatable and reimbursable by insurance companies multiplies (Kutchins & Kirk, 1997). On the other hand, the long-term effects created by the introduction of *DSM-III* in 1980 are more difficult to predict. The paradigm shift to the medical model marked by *DSM-III* might *increase* the number of articles published because mental disorders were now defined in terms that were more consistent with a medical approach (Kirk & Kutchins, 1992; Rogler, 1997). Similar to the increase based on the number of treatable illnesses, this would signal an increased need to update primary care physicians on new treatment options. Alternatively, this second dimension of medicalization might lead to a *decrease* in the number of articles published. In that case, a paradigm shift towards the medical model may lead physicians to treat mental disorders in a more routinized fashion. Thus, a reduction in articles on these topics would signal that mental health treatment is becoming incorporated into everyday family practice routines (McKinlay, 1982).

Drug Innovations

Drug therapy has revolutionized psychiatric treatment. Prior to 1950, only a few sedatives were available for treating mental disorders. However, within the last few decades, a broad range of drugs has become available for treating mental disorders. Since the late 1980s, several new classes of drugs have provided major breakthroughs in treating symptoms of mental distress. Buspirone (trade name Buspar) was marketed in 1987 to treat symptoms of anxiety; fluoxetine (trade name Prozac) was marketed in 1988 for major depression, and clozapine (trade name Clozaril) was marketed in 1990 for schizophrenia.[1]

These new classes of drugs were seen as more effective and better tolerated than existing drugs. They were thought to be especially beneficial for groups of patients not responsive to earlier drug treatment, unable to tolerate the side effects of earlier formulations, or at risk of dependency on other drugs (Marder & Van Putten, 1988; Schatzberg, Cole & DeBattista, 1997; Tierney, McPhee & Papadakis, 1997).

Because drugs make mental health problems more treatable within a primary care setting, new drugs should be of major interest to family practitioners. Indeed, drug education is a major component of CME. Journal editors claim their mission is to provide "balanced discussions of both the strength and weaknesses of diagnostic and treatment strategies" (AFP, 1999, 1). Thus, we should find an increase in the number of articles in years when new classes of drugs are introduced. Since two of the three new classes of drugs, Prozac and Clozaril, were designed specifically to treat major depression and schizophrenia, respectively, their impact should be most notable in SMI articles.

DATA AND METHODS

Our data set consists of all mental health articles published in two family practice journals, the *American Family Physician* (*AFP*) and the *Journal of Family Practice* (*JFP*) between 1977 and 1996 (*n* = 863). These are the only two U.S. family practice journals that were established in the 1970s and remain in publication today. Content analyses are based on articles in these journals that focus specifically on SMI during this time period (*n* = 202). These two specialty journals, whose professed purpose is to further medical education through "current updates on the diagnosis and treatment of clinical conditions" (American

Acadamy of Family Physicians, 1999, 1), have considerable influence among family practitioners. *AFP*, with 156,000 "qualified" subscribers (i.e. practitioners or medical students), is the official clinical journal of the American Academy of Family Physicians. The National Library of Medicine ranked *JFP*, with a qualified circulation of 75,000, highest in terms of impact. The Society of Teachers of Family Medicine ranked both *AFP* and *JFP* within the top three family practice journals for medical education (Miller, 1982).

Both journals were established in the 1970s and were the first major journals geared towards family practitioners that remain in publication today. Each publishes at least 12 times a year, and in 1993, *AFP* increased to 16 issues per year. *JFP* periodically publishes special issues, notably, the 1996 mental health supplement. These two journals also differ significantly. The editorial board of *JFP* includes both MDs and Ph.D.s, and articles often present the results of empirical research. The editorial board of *AFP* includes MDs only, and many articles provide specific directives on how to identify and treat illnesses.

We constructed our data set based on all original scientific articles and editorials that had a by-line and/or a brief description in the table of contents. Features, letters, or abstracts of research published elsewhere were excluded. Of the 6,259 articles fitting these criteria, 863 (14%) dealt with topics related to mental illness (see Table 1). We defined articles as related to mental illness if the title or a brief description in the table of contents specifically addressed a mental health issue or a diagnosable mental disorder as defined by the *DSM-IV*.[2] We use the *DSM-IV* to classify these 863 articles into four disorder-specific categories:

- *severe mental illnesses* (SMI), which includes the serious and persistent illnesses of schizophrenia, major depression and bipolar disorder;
- *other mental illnesses*, disorders such as anxiety, somatoform and personality disorders which, while less serious, still significantly compromise mental functioning;
- *behavioral disorders*, which includes e.g. substance abuse, eating and sexual disorders, plus other problem behaviors which interfere more with social than mental functioning; and
- *mental health issues*, a residual category of general, non-disorder specific articles on various mental health topics (i.e. training in psychiatric techniques and psychotropic medications in general, plus the relevance of psychiatric intervention by medical doctors).

To analyze trends in the amount of attention to mental illness, we first graph the number of articles over time. We then use pooled-time series analyses to investigate whether other concurrent changes in the mental health system

Table 1. Distribution of Articles on Mental Disorders in Family Practice
Journals, 1977–1996.

	Percent of MI articles $N = 863$	Percent of all articles $N = 6,259$
Severe Mental Illnesses (SMI) ($N = 202$)	23	3
Mood disorders ($N = 158$)	18	
Psychotic disorders ($N = 44$)	5	
Other Mental Illnesses ($N = 178$)	21	3
Anxiety disorders ($N = 54$)	6	
Somatoform disorders ($N = 53$)	6	
Dementia, delirium ($N = 40$)	5	
Disorders due to Med Cond. ($N = 13$)	2	
Personality disorders ($N = 11$)	1	
Misc. DSM disorders ($N = 7$)	1	
Behavioral Disorders ($N = 369$)	43	6
Substance Abuse ($N = 125$)	14	
Other Conditions ($N = 121$)	14	
Childhood disorders ($N = 47$)	5	
Sexual disorders ($N = 30$)	3	
Sleep disorders ($N-24$)	3	
Eating disorders ($N = 22$)	3	
Articles on General Mental Health Issues ($N=114$)*	13	2
Total	100	14

* Includes articles on non-disorder specific mental health topics.

(deinstitutionalization, the rise of managed care, revisions of the *DSM*, drug innovations) influence these trends.[3]

Our pooled time series models employ two dependent variables.[4] The main analysis is based on all 863 mental health articles. A second analysis uses only the 202 articles on depression, bipolar disorder and schizophrenia. This subset of SMI articles is also the basis for our more detailed content analysis.[5] All models contain controls for the total number of articles published each year, a dichotomous variable indicating each journal, a second dummy variable for the 1996 *JFP* special issue on mental health, and a third dummy control capturing changes in journal editors (to control for any possible biases in editorial decision making).

We measure deinstitutionalization as yearly counts of the number of state mental health beds.[6] We measure managed care as yearly counts of HMO enrollments. Possible *DSM* effects are measured using two separate variables: Since the paradigm shift from a psychodynamic model to a medical model occurred in the third edition published in 1980, we use a dummy variable coded 1 for the year this shift took place and all years thereafter. A second variable measures the more continuous effect of revisions, counting the changing number of codes with each edition of the *DSM*.[7] A dichotomous variable serves to measure major drug innovations, coded 1 for any year in which a major new class of drugs was released.[8]

Finally, content analyses of SMI articles ($n = 202$) look more closely at the influence of drug innovations and *DSM* revisions on information for family doctors. We coded these articles with regard to whether they recommended drugs or psychosocial treatment as the first line of treatment and whether they discussed side effects. One researcher coded all the articles while a second read and resolved the ambiguous or problematic cases. 10% of the articles were coded independently and inter-rater reliability computed. Agreement between the two coders was high with values ranging from 86 to 95% (kappa values ranged from 0.75 to 0.92).

RESULTS

Distribution of Mental Health Articles

Table 1 illustrates the distribution of articles across categories. Behavioral disorders, including diverse topics of substance abuse, sexual and eating disorders, constitute the largest grouping of articles (43% of all mental illness articles).[9] Given that behavioral disorders have the fewest medical treatment indications, we were surprised by the amount of attention journals devoted to them (especially those under the category of "other conditions," such as emotional problems surrounding illness, marital problems, physical and sexual abuse and life transitions). This focus may reflect the somewhat holistic nature of family practice and its commitment to broad comprehensive care for the entire family (American Board of Medical Specialties, 1999). In contrast, articles on SMI, purported to be those with the strongest medical underpinnings, comprise 23% of total mental illness articles and 3% of all articles in these journals. These patterns are especially interesting because general practitioners play such a key role in treating disorders classified as SMI.

Figure 1 shows the trend in the percentage of journal articles devoted to mental health. The top line of the figure represents the combined percentage for all mental health articles; the lower lines indicate the percentage of articles within each subcategory. The percentage of articles devoted to mental health ranges from a high of 21% in 1987 to a low of not quite 10% in 1985–1986 and 1993–1994. While there is little indication of a consistent increase or decrease in attention to mental health over this period, the fluctuations suggest that there may be multiple processes influencing the number of articles. For example, it appears that there are at least two distinct periods of change over these two decades. During the early to mid 1980s there was a gradual decline in attention to mental illness, with articles on mental health dropping from 18 to 10% of total articles. This decline is followed by renewed attention to mental health in 1987. Interestingly, this peak is then followed by another period of gradual decline that continues through 1996.

Patterns for each mental health subcategory (SMI, other mental illness, behavioral disorders and mental health issues) largely follow that found for all articles. This suggests that overall trends were not disproportionately influenced by attention to any single subcategory. The one exception is a steady decline in the amount of attention to general "mental health issues." In the early 1980s, the percentage of articles on mental health issues (which in the late 1970s and early 1980s frequently dealt with training in mental health care) matched or exceeded that for either behavioral or SMI disorders. By the early 1990s, the percentage of articles on behavioral disorders far exceeds that for other categories, and articles on general mental health issues have largely disappeared.

Our time series analyses explore how multiple countervailing factors produce these fluctuations in general mental illness and SMI articles. Table 2 lists data sources and descriptive statistics of variables included in the analyses. Table 3 presents the time series results. Results suggest that, after controlling for likely influences from journal type, the total number of articles published, changes in journal editor and the *JFP* special mental health supplement, the strongest influence comes from drug innovations (model 2). In other words, the amount of coverage family practice journals give mental illness increases significantly in years when major new classes of drugs are introduced.

Additional models (not shown) provide further evidence that expanding pharmaceutical technology may be the driving force for increased attention, rather than the influence of any particular drug. For instance, we also examined the impact of individual drugs by employing separate dummy variables for each of the three breakthrough drugs and found that the first breakthrough drug introduced (the anti-anxiety drug Buspar) had the strongest influence on the number of articles. Interestingly, the more widely advertised and prescribed drug, Prozac, which was introduced one year after Buspar, has a far weaker influence on the number of mental illness articles. This suggests that major pharmaceutical innovations renew attention to mental health treatment in family practice journals. Our subsequent analyses of the content of the articles will

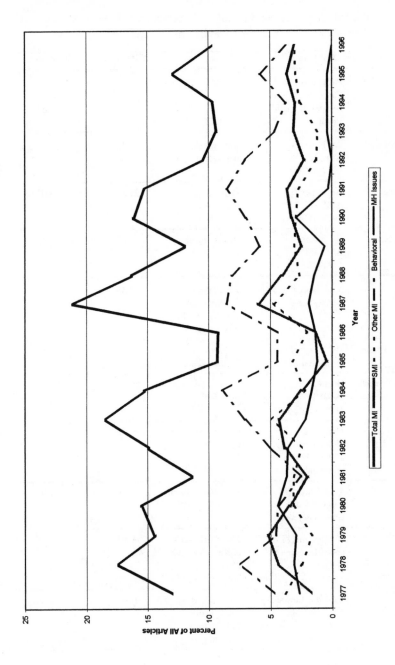

Fig. 1. Percent of Family Practice Journal Articles on Mental Health Illness, Serious Mental Illness, Behavioral and other Disorders, and Mental Health Issues, 1977–1996.

Table 2. Pooled Time Series: Variables, Data Sources, Metric, Means and Standard Deviations.

Variable	Data Source	Metric	Mean	Stand. Dev.
Journal	American Family Physician; Journal of Family Practice	1 = AFP 2 = JFP	1.50	0.51
Total Articles	All issues: AFP, JFP 1977–1996	Raw number per year	156.48	35.87
Mental Illness Articles	All issues: AFP, JFP 1977–1996	Raw number per year	21.58	9.03
SMI Articles	All issues: AFP, JFP 1977–96	Raw number per year	5.05	2.72
Editor	Year journal editor changed	1 = 1989(AFP) 0 other years 1 = 1991 (JFP) 0 other years	0.35	0.48
JFP Special Issue	1996 special issue on mental health	1=1996 0 other years	0.03	0.16
HMO Enrollments	Health, U.S.: 1987, 1990 1998, interpolated	number per year (in 1,000)	25.37	14.91
State Hospital Mental Health Beds	Health, U.S.: 1982, 1986, 1992, 1995, 1998, interpolated	number per year (in 1,000)	122.01	35.51
DSM-III+	Year DSM-III published and all subsequent years	1 = 1980 – 1996 0 = 1977 – 1979	0.85	0.36
DSM-count	Number of DSM codes by edition of DSM	217 = 1977–1979 233 = 1980–1986 256 = 1987–1993 366 = 1994–1996	258.60	47.73
New Class of Drugs	Year breakthrough drug marketed	1 = 1987 (Buspar), 1988 (Prozac), 1990 (Clozaril) 0 other years	0.15	0.36

look further at this issue by examining shifts in topics and information about these drugs.[10]

The two indicators of the changing supply of mental health patients to general practitioners have smaller influences on mental health coverage in these journals. As expected, both HMO enrollments and the number of state mental health beds indicate that changes in the supply of mental health patients directly influence the number of articles on mental health. The trend towards fewer state hospital beds indicates a greater demand for more community care, thus increasing the need for information about mental health treatment. This increase is offset, however, by restrictions on care for patients with chronic mental illness and the expansion of special "carve out" programs (Bennett, 1988; Bittker, 1985; McFarland, 1994). Thus, despite the increasing supply of patients generated by deinstitutionalization, rising HMO enrollments tend to reduce the proportion of patients with severe mental illness treated by family practitioners. This, in turn, leads to a decrease mental illness information provided to family practitioners.[11]

We now turn to the complex effects the medicalization of mental illness has had on the number of articles published. First, the expansion of the number of diagnostic codes in each edition of the *DSM* codes increased attention to mental health and illness, though its influence remains modest. This suggests that the increase in the number of formal diagnostic codes did have some influence on editorial decisions regarding mental health articles. Many of the articles were authored by mental health professionals and provided updates on symptom recognition as well as treatment. Thus as the number of treatable conditions increased, the number of articles identifying these conditions also increased.

Second, the paradigm shift toward the medical model of mental illness marked by the adoption of the *DSM-III* appears to have *decreased* the amount of attention given to mental illness. The drop in the number of articles after the introduction of the *DSM-III* may indicate that success in shifting the definition of mental disorders from psychological to more easily treatable "medical" problems routinized mental health treatment for general practitioners. In the paragraphs below we examine this argument in more detail by looking at the content of SMI articles.

Given that drug innovations produced the strongest empirical effect on the coverage of mental health issues in general, the question is whether changes in the number of SMI articles published paralleled the trends discussed above. Models 3 and 4 show that factors influencing increases in the number SMI articles ($n = 202$) are indeed very similar to those influencing all mental health articles. This further underscores the importance of conducting a content analysis of the SMI articles. We now focus on the subsample of SMI articles to explore

Table 3.   Pooled Time Series Analysis of Trends in All Articles on Mental
Illness and Articles on Severe Mental Illness in Two Family Practice
Journals, 1977–1996 (Standard errors in parentheses).

	ALL ARTICLES		SMI ARTICLES	
Variable	MODEL 1 Base Model	MODEL 2 Full Model	MODEL 3 Base Model	MODEL 4 Full Model
Total articles	0.20***	0.20***	0.03**	0.03***
	(0.02)	(0.02)	(0.01)	(0.01)
Journal	−2.49	−2.68*	0.46	0.18
	(1.58)	(1.05)	(0.88)	(0.65)
Journal editor	−4.16**	−1.82	−0.71	−0.42
	(1.54)	(2.10)	(0.84)	(1.21)
JFP Special Issue	1.07	9.92*	2.27	5.97**
	(5.42)	(4.70)	(2.48)	(2.29)
HMO Enrolled		−0.59***		−0.30***
(x 1000)		(0.17)		(0.09)
State MH Beds		−0.23**		−0.13***
(x 1000)		(0.08)		(0.04)
DSM-III+		−10.26**		−6.24***
		(3.63)		(1.87)
DSM-count		0.04ᵗ		0.02*
		(0.21)		(0.01)
New Class		9.43***		3.15**
		(1.95)		(0.99)
Constant	−4.3128	34.33*	−0.56	21.21
	(3.57)	(15.82)	(1.93)	(8.06)
Log Likelihood	−121.146	−109.25	−90.06	−80.59
Chi-Square	105.85***	288.89***	13.05*	49.62***
N	40	40	40	40

$^t p \leq 0.10$, $* p \leq 0.05$, $** p \leq 0.01$, $*** p \leq 0.001$, two-tailed t-test,
Models corrected for first-order autocorrelation.

more closely how pharmaceutical innovations or the adoption of the *DSM*
influence their content.

The Influence of Drug Innovations on Article Content

Innovation in drug therapy clearly has a powerful effect on the amount of attention journals give to mental illness. This confirms the journals' stated purpose
to provide continuing education to family practitioners. But do drug innovations alter the *content* of SMI articles? Specifically, do articles tend to stress
drug treatment as the primary source of therapy as major new drugs are intro-

duced? Does the introduction of new drugs increase the amount of discussion of side effects associated either with the new formulations or older drugs replaced by the newer ones?

Figure 2 examines long-term shifts among SMI articles recommending drugs versus psychosocial treatment.[12] Of particular interest are the three years in which major innovative drugs were introduced (Buspar in 1987, Prozac in 1988 and Clozaril in 1990). The long-term trends show clearly that drug treatment is the dominant recommendation, consistent across the entire twenty-year time period examined. At any time, between 60% and 80% of the articles recommend drugs as the treatment of first resort. Interestingly, though based on a relatively small number of articles ($n = 202$), it appears that the percentage of articles that recommend drugs as the primary treatment declines slightly during this period of drug innovation. Conversely, recommendations for psychosocial treatment rise again in the mid-1990s. Thus, while drug innovations increased the amount of attention to mental illness, the content of the articles published did not shift appreciably in favor of pharmaceutical therapy, and even appear to have reduced primary recommendations of drug treatment for a period.

Given that the treatment options discussed generally remain quite stable over time, perhaps the content of the articles changed regarding the discussion of side effects? Figure 3 examines this question. The top line repeats the percentage of articles recommending drug treatment shown in Fig. 2; the bottom line represents those articles discussing side effects. If the slight decline in drug treatment recommendations evident in the early 1990s stems from concern over side effects, we should find increased discussion of side effects. In fact, beginning in the late 1980s, declines in articles recommending drugs as the first line of treatment are matched by a steady increase in articles discussing side effects. However, these patterns are not particularly clear-cut.

One explanation for this finding is that the introduction of new drugs actually boosts attention to the problematic side effects of psychotropic drugs. While the tendency to produce fewer side effects than existing remedies is one of the new drugs' most salable features, discussion of the new drug's merits often necessitates revisiting the old drug's shortcomings (Valenstein, 1998). Additionally, two of the three new classes of drugs (Prozac & Clozaril) had potentially serious side effects in their own right that even garnered the attention of the popular press: Prozac was suspected of making patients suicidal or violent (Breggin & Breggin, 1994), and the market release of Clozaril was delayed because it carried a life-threatening side effect called agranulocytosis. While the charges against Prozac were not substantiated, the FDA approved Clozaril only under the condition that patients undergo weekly blood screenings for the presence of this rare and potentially fatal blood disorder (Marder & Van Putten, 1988).

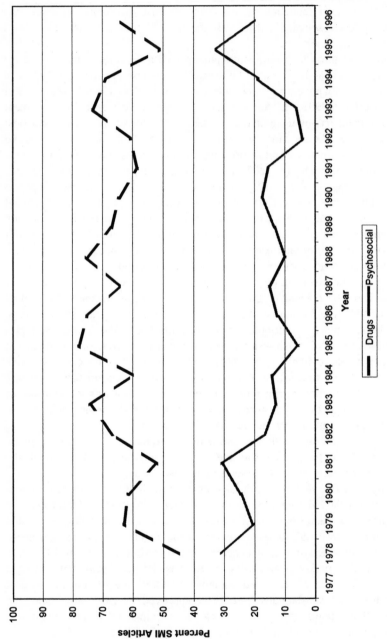

Fig. 2. Percent of Articles on Severe Mental Illness Recommending Drugs or Psychosocial Therapy as Primary Mode of Treatment, 1977–1966 (2 Year Moving Average).

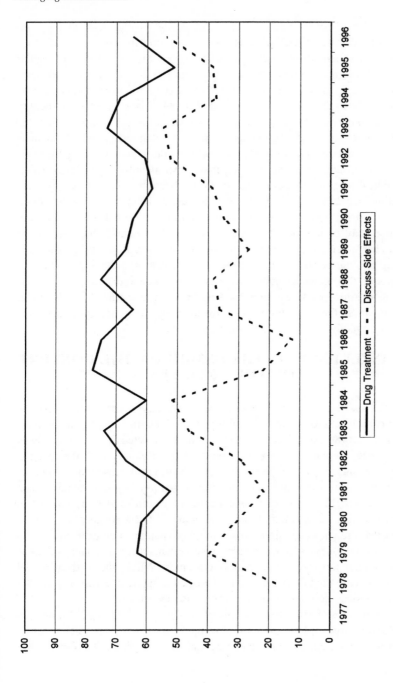

Fig. 3. Percent of Articles on Severe Mental Illness Recommending Drugs as Primary Mode of Treatment with Percent With at Least One Paragraph Discussion of Side Effects, 1977–1996 (2 Year Moving Average).

Do the slight 1990s trends towards discussing psychosocial therapy and wariness of side effects spell dissatisfaction with drug therapy and a call for a return to earlier methods of treatment such as psychotherapy? A careful reading of the articles written during this time suggests this is not the case. Rather it appears that drug treatments have become utterly normative, requiring little discussion. These articles do not suggest that physicians avoid drug therapy, but rather *assume* drug therapy, but remind physicians to not completely neglect the therapeutic value of a good doctor-patient relationship, which includes talking to the patient. For example, one article points out that drug treatment is necessary but not sufficient, warning that "when depression is viewed solely as a medical disease to be treated with medication, relevant psychosocial contributors may be overlooked" (Shearer & Adams, 1992, 435). Another article points out that "(r)ecognition, watchful waiting, informal counseling and negotiation may be appropriate strategies for some patients" (Sussman, 1995, 541). Rather than suggesting that treatment for mental health problems is becoming less medicalized, these articles suggest that drug therapy has become so *dominant* that general practitioners need reminders that important other modes of treatment exist. So embedded is the medical model in the treatment of mental illness that assumptions of drug treatment have become routine and thus are less likely to merit close attention in these articles.

THE INFLUENCE OF THE DSM-III ON THE CONTENT OF THE ARTICLES

Time series analyses also revealed a significant decrease in the amount of coverage of mental health topics in the family practice literature after the adoption of the *DSM-III* in 1980. At first glance, we might have expected this shift towards the medical model to place problematic behaviors more clearly with the realm of medical care, thus increasing interest in mental health problems in these journals. Indeed, trends in recommendations for drug versus psychosocial therapy in the early 1980s (see Fig. 2) suggest that treatment recommendations did indeed become more medicalized after the adoption of the *DSM-III*. An alternative explanation is that the paradigm shift from a psychodynamic framework toward the view that mental illness is "a disease like any other" resituated mental health treatment in familiar territory for general practitioners. Within the medical model, mental health care shed its esoteric and time-consuming treatment strategies (that required significant journal attention) for routine medical care.

A closer look at SMI articles validates this view. A 1978 *JFP* article on schizophrenia urges physicians to develop a "long-term physician-patient relationship" with schizophrenic patients, and to allow them "to discuss fears and anxieties openly" (Donlon, 1978, 75). Although descriptions of treatment options with antipsychotic drugs are mentioned, the thrust of the article is that the family doctor is responsible for overseeing comprehensive treatment of the patient through all phases of the illness, including vocational and rehabilitative services in the community. A similar *JFP* article (Scher et al., 1980) acknowledges the difficulties a physician may encounter with schizophrenic patients. The article builds on the premise that drug treatment is necessary but not sufficient and seeks to provide practical suggestions for the day-to-day management of such patients. Difficulties in communicating with patients and the importance of guiding and educating both the patient and family are discussed at length (Scher et al., 1980, 411).

However, by 1983, an article published in the same journal reflects a more simplified and narrow view of the family practitioner's role in treating schizophrenia (Katon & Ries, 1983). Diagnostic criteria, cited from *DSM-III*, are described as "comparable to medical diagnosis" (p. 100), and the focus of the article is on the importance of drug therapy and the physician's role in finding the right dosage and being mindful of side effects. Whereas the good doctor-patient relationship was considered paramount for maintaining mental health in 1978, by 1983, the emphasis in the family practice literature had dramatically shifted to view the doctor-patient relationship as critical in order to ensure patient compliance with drug treatment.

This shift, which we also found for discussions of depression in both journals, suggests that medicalization, as indicated by the adoption of *DSM-III* and the amount and focus of discussion of treatment options, significantly simplified the physician's role in treating mental disorder. Early mental health care, which once required the family practitioner to act as therapist, social service coordinator and physician, became more focused over time upon managing and monitoring pharmaceutical treatments.

DISCUSSION

Medical journals serve as an important source of continuing medical education for physicians. Updates on advances in technology, most notably in terms of drug innovations, and shifts in approaches to treatment are especially crucial for primary care providers, who routinely confront the widest range of problems in day to day practice. Since general practitioners have begun to provide mental health care in the course of primary care, our focus on the amount and content of information in family practice journals provides one important lens for viewing what researchers have called the "de facto mental health system" (Regier et al., 1978). We are interested both in the amount of mental health information provided to family practitioners over time, and in the degree to which structural and treatment changes in the mental health system influence journal attention to mental illness.

When analyzing trends in the amount of information, we assumed that increases in the number of articles on mental illness reflect greater interest (at least by editors and authors) in how to best treat patients identified as having mental health problems. As expected, we found that broader structural changes altering the supply of patients to general practitioners influenced trends in publications. While deinstitutionalization increased journal attention to mental illness, the growth of managed care, which in some cases tended to divert patients to other providers, reduced the amount of attention. This suggests that journals are fairly responsive to the mix of patients seen by practitioners, but it also shows how structural changes in the financing of health care can even influence health care information provided to specialists.

More importantly, quantitative analyses revealed that drug innovations provided the strongest impetus for increased coverage of mental health care. However, we found this attention was generated less by the introduction of any specific drug than by general change in pharmaceutical technology. Neither was the increased coverage confined to the discussion of the illnesses for which these drugs had been developed. Instead, drug innovation renewed attention to all types of mental illness. This suggests that the influence of drug innovation goes beyond its immediate discussion, thus pointing to the broader power of innovation to frame the entire context in which mental illness is defined and treated.

A closer look at the content of articles also suggests a similar pattern. We found that the pharmaceutical innovations had a broad influence on treatment recommendations. Further research might explore the content of the articles to determine the impact of widely prescribed drugs like Prozac on the medical literature. Further research might also explore the impact of other sources of information on mental health care provided to family practitioners such as journal advertising or drug company representatives.

Finally, analyses indicate that the formal shift towards a medical model embodied by the introduction of *DSM-III* shaped both the amount and the content of mental illness articles published. On one hand, the long-term effects of the paradigm shift to the medical model led to a decrease in number of mental health articles published. On the other hand, changes in the content of SMI articles further illustrate the complexity of the routinization process. While drug treatment remained the primary recommendation throughout this period and even increased in the early 1980s, the paradigm shift with *DSM-III* had far broader consequences: It fundamentally redefined the role of family practitioners, narrowing it from multifaceted treatment provider to medications manager. By the early 1990s, this role had become so established that articles began to remind practitioners to not forget the importance of interacting with their patients.

Attention to both the amount and content of mental health information also points to the interplay between processes of medicalization and routinization. In the process of routinization, topics begin to be seen as familiar, requiring little further investigation (Schutz, 1970). To the extent that medicalization streamlines and narrows the recommended treatment of mental illnesses, routinization may constitute an important later stage of medicalization (McKinlay, 1982). Once routinization occurs, the treatment of mental disorders is presented as less problematic, and treatment information is only published when new innovations occur. In this context, trends in the late 1980s, during and after three major drug innovations, are particularly informative. The renewed attention to mental health treatment points to the power of pharmaceutical innovations to revitalize ongoing debates, potentially even taking medicalization to new levels. Thus, we suggest that future research on medicalization consider the role routinization plays in different stages of the medicalization process.

To summarize, several decades of research have documented the medicalization of mental disorders (Brown, 1995; Conrad, 1975, 1992; Conrad & Schneider, 1980). So far, most research has focused on the efforts of psychiatry to change definitions of mental disorders (Conrad, 1975; Figert, 1996; Kutchins & Kirk, 1997; Scott, 1990). But the full medicalization of mental disorders also depends on other medical specialties accepting these definitions and incorporating the treatment of mental problems within their domain of care. Moreover, the medicalization of mental illness has also been accompanied by dramatic shifts in the philosophical, institutional, and financial aspects surrounding mental health care. Because researchers have paid relatively little attention to the increasing importance of care provided by non-psychiatric clinicians, our article focuses on continuing mental health education of general practitioners, who today constitute the "de facto mental health system". We hope our findings prompt further consideration of other aspects of mental health care provided by general practitioners, particularly the nature of that care, the attitudes of practitioners towards patients, and how both may have changed over time.

ACKNOWLEDGMENTS

This research has been supported by a fellowship from Indiana University to the first author and by a grant from the National Institute of Mental Health

(R24-MH51669). We thank Brian Powell and Cathy Johnson for their comments and suggestions.

NOTES

1. We classified drugs as belonging to a "new class" by their position in broad categorical groupings. For example, Clozaril belongs to a class of "atypical" anti-psychotics that control psychosis with fewer extra-pyramidal side effects than earlier fomulations. Similarly, Buspar is unrelated to benzodiazepines which also treat symptoms of anxiety, and Prozac is a selective serotonin reuptake inhibitor (SSRI), not a tricyclic anti-depressent like its predecessors.

2. We chose DSM-IV as a standard because it is the most comprehensive edition to date. As the DSM has expanded in successive editions, all conditions identified in the twenty-year period were included.

3. Analyses used the "xtgls" function in STATA, which is designed for data sets containing more time points (years) than cross-sectional units (journals). It employs feasible generalized least squares and allows us to estimate models despite autocorrelation and heteroskedasticity. Heteroskedasticity was not a problem, but reported results are corrected for first-order autocorrelation. Models without AR(1) corrections were also estimated but are not shown. Estimates for these models were substantively identical, though significance levels of some of the variables were slightly lower and the Wald χ^2 reported suffered noticeably when we did not control for autocorrelation. We also estimated models using panel- (journal-) specific autocorrelation corrections, which yielded the same results as those presented here.

4. Analyses (not shown) also included estimates with other dependent variables (e.g. separate analyses of mood disorders and psychotic disorders). Results for the subgroups were volatile because of the small number of articles in these subcategories. We also ran regressions for each journal separately, but the few degrees of freedom (df = 12) made results volatile. Despite this volatility, the direction and order of strongest effects remained similar to those shown.

5. Descriptive statistics revealed no need to log for the dependent or other variables to correct for skewness. Similarly, we estimated all models using 1-year lags for independent variables. None of these measures produced stable, statistically significant results, indicating that drug and *DSM* innovations as well as changes in the supply of patients caused by deinstitutionalization and HMO growth were quickly reflected in these journals.

6. Alternative deinstitutionalization measures included counts of long-term state psychiatric hospital beds derived from different censuses (American Hospital Association; National Institute of Mental Health) as well as the number of state mental hospitals. All three measures were correlated at 0.95 or above. The measure we use produced the best model fit and has the advantage that it takes into account short- and long-term placement facilities, thus providing a more fine-grained indicator of supply than the other measures. We found substantively identical effects regardless of the deinstitutionalization measure used.

7. In analyses not reported here, effects of *DSM* versions were estimated individually and in various combinations, using dichotomous measures. High correlations prevent including all versions simultaneously. The last two *DSM* versions have the strongest

influence on the number of articles published, though the substantive results were similar for any combination of *DSM* revisions. We considered the possibility that publication delays may cause lagged rather than immediate effects on the number of mental health articles. When tested, models lagged by one year produced non-significant results for the *DSM* and other measures, making the use of lags empirically and theoretically questionable.

8. Alternative specifications of drug innovations include separate dummy variables for each of the three breakthrough drugs, Buspar, Prozac and Clozaril. Results not reported here showed the combination measure to have the strongest, most consistent effects on articles published.

9. In time series analyses not reported here, we tested the effects of these structural changes on the number of articles dealing with behavioral disorders. Coverage of behavioral illnesses strongly depended on the journal. Deinstitutionalization increased the number of articles, but the shift to the medical model captured by *DSM-III* had a negative effect. Neither the increase of problems defined as medically treatable, nor drug innovations, nor the growth in managed care had any empirical effect on articles dealing with behavioral disorders.

10. A broader measure reflecting the number of all new psychotropic drugs introduced each year (not just the breakthrough drugs) was not related to increases in the number of articles.

11. Additional analyses (not shown) checked for the possibility of multicollinearity between HMO enrollments and the number of state mental health beds. The correlation between these two variables is low and entering each variable separately in the models did not change the results.

12. Possible categories of treatment included drugs, electroshock, hospitalization, psychological "talk" therapy and social treatments, such as increasing family or network supports. Our analyses combine these last two categories into a single "psychosocial" category to contrast this more socially-based approaches to medical treatment. Drug treatment drew the vast majority of recommendations among the medical treatment options. Many articles recommend multiple strategies, and in these cases, decisions about which was the primary treatment recommended were based upon the amount of attention given to each approach.

REFERENCES

American Academy of Family Physicians (1999). *American Family Physician*: Circulation Information: http://www.aafp.org/afp/circulat.html

American Board of Medical Specialties. (1998). *The Official ABMS Directory of Board Certified Medical Specialists*. New Providence, NJ: Marquis Who's Who.

American Psychiatric Association (1968). *Diagnosis and Statistical Manual of Mental Disorder* (2nd ed.). Washington, D.C.: American Psychiatric Association.

American Psychiatric Association (1980). *Diagnosis and Statistical Manual of Mental Disorder* (3rd ed.). Washington, D.C.: American Psychiatric Association.

American Psychiatric Association (1987). *Diagnosis and Statistical Manual of Mental Disorder* (3rd ed.). Washington, D.C.: American Psychiatric Association.

American Psychiatric Association (1994). *Diagnosis and Statistical Manual of Mental Disorder* (4th ed.). Washington, D.C.: American Psychiatric Association.

Bennett, M. J. (1988). The Greening of the HMO: Implications for Prepaid Psychiatry. *American Journal of Psychiatry*, *145*(12), 1544–1549.

Bittker, T. E. (1985). The Industrialization of American Psychiatry. *American Journal of Psychiatry*, *145*(2), 149–153.

Bittker, T. E. (1992). The Emergence of Prepaid Psychiatry. In: J. Feldman & R. J. Fitzpatrick (Eds), *Managed Mental Health Care: Administrative and Clinical Issues*. Washington D.C.: APA Press.

Breggin, P. R., & Breggin, G. R. (1994). *Talking Back to Prozac*. New York: St. Martin Press.

Brown, P. (1985). *The Transfer of Care*. London: Routledge.

Brown, P. (1995). Naming and Framing: The Social Construction of Diagnosis and Illness.*Journal of Health and Social Behavior*, (Extra Issue), 34–52.

Conrad, P. (1975). The Discovery of Hyperkinesis: Notes on the Medicalization of Deviant Behavior. *Social Problems*, *23*, 12–21.

Conrad, P. (1992). Medicalization and Social Control. *American Review of Sociology*, *18*, 209–232.

Conrad, P., & Schneider, J. (1980). *Deviance and Medicalization: From Badness to Sickness*. St. Louis: Mosby.

Cook, J. A., & Wright, E. R. (1995). Medical Sociology and the Study of Severe Mental Illness: Reflections on Past Accomplishments and Directions for Future Research. *Journal of Health and Social Behavior*, (Extra Issue), 95–114.

Donlon, P. T. (1978). The Schizophrenias: Medical Diagnosis and Treatment by the Family Practitioner. *Journal of Family Practice*, *6*(1), 71–82.

Falloon, I. R., & Fadden, G. (1993). *Integrated Mental Health Care*. Cambridge: Cambridge University Press.

Figert, A. E. (1996). *Women and the Ownership of PMS: The Structure of a Psychiatric Disorder*. New York: Aldine De Gruyter.

Goldberg, D., & Huxley, P. (1980). *Mental Illness in the Community: The Pathway to Psychiatric Care*. London: Tavistock Publications.

Grob, G. (1991). *From Asylum to Community: Mental Health Policy in Modern America*. Princeton, NJ: Princeton University Press.

Healy, D. (1997). *The Antidepressant Era*. Cambridge, MA: Harvard University Press.

Katon, W., & Reis, R. (1983). Schizophrenia. *Journal of Family Practice*, *17*(1), 99–102, 107–108, 111–114.

Kirk, S. A., & Kutchins, H. (1992). *The Selling of DSM: The Rhetoric of Science in Psychiatry*. NewYork: Aldine de Gruyter.

Kutchins, H., & Kirk, S. A. (1997). *Making Us Crazy: DSM: The Psychiatric Bible and the Creation of Mental Disorders*. New York: The Free Press.

Marder, S. R., & Van Putten, T. (1988). Who Should Receive Clozapine? *Archives of General Psychiatry*, *45*, 865–867.

Mayer, T. R., & Mayer, G. G. (1985). HMO's: Origins and Development. *The New England Journal of Medicine*, *312*(9), 590–594.

McFarland, B. H. (1994). Health Maintenance Organizations and Persons with Severe Mental Illness. *Community Mental Health Journal*, *30*(3), 221–242.

McKinlay, J. (1982). From "promising report" to "standard procedure:" Seven stages in medical innovation. In: J. B.Milbank (Ed.), *Technology and the Future of Health Care. Milbank Reader, 8*, 233–270. Cambridge, MA: MIT Press.

Mechanic, D. (1989). *Mental Health and Social Policy* (3rd ed.). Englewood Cliffs, NJ: Prentice-Hall.

Miller, M. D. (1982). Ratings of Medical Journals by Family Physician Educators. *Journal of Family Practice*, *15*(3), 517–518.

Miranda, J. (Ed.), (1994). *Mental Disorders in Primary Care*. San Francisco: Jossey-Bass Publishers.

Orleans, C. T., George, L. K., Houpt, J. L., & Brodie, H. K. (1985). How Primary Care Physicians Treat Psychiatric Disorders: A National Survey of Family Practitioners. *American Journal of Psychiatry, 142*(1), 52–57.

Regier, D. A., Goldberg, I. D. & Taube, C. A. (1978). The De Facto U.S. Mental Health System. *Archives of General Psychiatry, 35*, 685–693.

Rogler, L. H. (1997). Making Sense of the Historical Changes in the Diagnostic and Statistical Manual of Mental Disorders: Five Propositions. *Journal of Health and Social Behavior, 38*(1), 9–20.

Sartorius, N., Goldberg, D., deGirolamo, G., Costa e Silva, J. A., Lucrubier, Y., & Wittchen, J. U. (Eds) (1990). *Psychological Disorders in General Medical Settings*. Toronto: Hogrefe & Huber, Publishers.

Schatzberg, A., Cole, J. O., & DeBattista, C. (1997). *Manual of Clinical Psychopharmacology*, (3rd ed.). Washington, D.C.: American Psychiatric Press.

Scher, M., Wilson, L. & Mason, J. (1980). The Management of Chronic Schizophrenia. *Journal of Family Practice, 11*(3), 407–413.

Schurman, R. A., Kramer, P. D., & Mitchell, J. B. (1985). The Hidden Mental Health Network. *Archives of General Psychiatry, 42*, 89–94.

Schutz, A. (1970). *Reflections on the Problem of Relevance*. New Haven: Yale University Press.

Scott, W. J. (1990). PTSD in DSM-III: A Case of Politics of Diagnosis and Disease. *Social Problems, 37*, 294–310.

Shearer, S. L., & Adams, G. K. (1993). Nonpharmacologic Aids in the Treatment of Depression. *American Family Physician, 47*(2), 435–441.

Starr, P. (1982). *The Social Transformation of American Medicine*. New York: Basic.

Sussman, J. I. (1995). Mental Health Problems in Primary Care: Shooting First and Then Asking Questions? *Journal of Family Practice, 41*(6), 540–542.

Tierney, L. M., McPhee, S. J., & Papadakis, M. A. (1997). *Current Medical Diagnosis and Treatment* (36th ed.). Stamford, CT: Appleton and Lange.

U.S. Bureau of the Census. *Statistical Abstract of the United States: 1980* (101st ed.); 1987 (107th ed.); 1998 (118th ed.). Washington, D.C. (1979; 1986; 1997).

Valenstein, E. S. (1998). *Blaming the Brain: The Truth About Drugs and Mental Health*. New York: Free Press.

Wegner, E. L. (1990). Deinstitutionalization and Community-Based Care for the Chronically Mentally Ill. *Research in Community Mental Health, 6*, 295–323.

PART IV:
ISSUES OF CHANGES FOR
SPECIALIZED PATIENTS: PEOPLE
WITH SERIOUS ILLNESSES

USING ACTION RESEARCH TO IMPROVE COLLABORATION BETWEEN BREAST CANCER PATIENTS AND PHYSICIANS: CREATING THE PROGRAM FOR COLLABORATIVE CARE

Caryn Aviv, Karen Sepucha, Jeff Belkora and Laura Esserman

INTRODUCTION

Recently, breast cancer patients have become more eager to actively participate in decisions about their care. The availability of information on the Internet, and the visibility of the breast cancer advocacy movement (Klawiter, 1999; Potts, 1999), have changed the way breast cancer is addressed in the United States. Most of the exchange between breast cancer patients and physicians occurs during the medical consultation. However, the decision making in breast cancer has yet to incorporate these broader social changes (Fisher, 1996). In the short time of a consultation, physicians disclose information about the diagnosis, explain alternatives for treatment and present information about the risks and benefits. Patients voice their preferences and disclose information

Changing Consumers and Changing Technology in Health Care and Health Care Delivery,
Volume 19, pages 207–229.

about their history. Somehow, during these rushed, strained conversations, patients and physicians make decisions about treatment.

This paper describes an action research project that aimed to improve the way patients and physicians make decisions about breast cancer. We describe: (1) the current problems breast cancer patients and physicians face as they make decisions about treatment, (2) the process of developing methods to improve the collaboration between patients and physicians, and (3) the transition from action research to implementing a full time clinical program. We use an ethnographic, case study approach to illustrate the social world of breast cancer, to highlight the difficulties in decision-making and to describe the process of conducting action research in a university medical center.

Action research is an interdisciplinary methodology with similarities to participatory sociological research (Hart & Bond, 1999; Cancian, 1996; McCormick, 2000). Action research emphasizes the formulation, implementation and assessment of interventions in applied settings, and advocates finding a motivated organizational partner who is willing and able to change the status quo. The partnership seeks to render an explanatory account of a problem, invent, practice and implement productive change, and then assess the effectiveness of the intervention. Success of an intervention is determined by the adoption and use of the methods (i.e. is it compelling and practical enough to adopt?). However, biomedicine stipulates that only statistical significance through randomized clinical trials can prove associations between variables, and therefore generate legitimate "results." To address this reality, we used both qualitative methods (ethnographic participant observation, focus groups, in-depth interviews) and a sequential, controlled clinical trial to determine the value of our interventions (Sepucha, Belkora et al., 2000).

Background

We found a willing action research partner in the University of California, San Francisco Breast Care Center (BCC). The director, a practicing surgeon, wanted to explore new tools and methods to improve the decision-making interaction and the quality of decisions. We used action research (Lewin, 1951; Hart, 1999) and its foundations in critical theory (White, 1995) as a guide for generating the important questions, as well as answering them. Although we have been involved in many projects – collaborations in classrooms, rape crisis centers, community resource centers, and other hospital units – this project was the first to focus on physicians and patients at the UCSF Breast Care Center from beginning to end.We focus on the portion of our collaboration that resulted in the intervention called Consultation Recording. In Consultation Recording, a trained researcher facilitates and records the medical consultation. This intervention extends and improves on another intervention called Consultation Planning, also developed by the authors. In Consultation Planning, a trained researcher helps patients prepare for an upcoming medical consultation by organizing their questions and concerns on a flow chart. The Consultation Plan acts as a visual aid or "road map" for the discussion between patients and physicians.

Organizational Context: The Breast Care Center

The Breast Care Center (BCC) is a busy, multidisciplinary clinic that opened in 1996. At first it was simply several private practices co-located in a cramped space. Scheduling was a challenge, exam rooms were always double booked, patients had to wait, and physicians seemed to get in each other's way more than they collaborated. The staff and services have since integrated, they have moved into new space that better supports patient care. The ten physicians and three nurse practitioners at the BCC see an average of 9200 patients a year from diverse communities across Northern California.

Every day at the Breast Care Center, patients and their families deal with a potentially life threatening disease and make decisions that have significant impact on their bodies and lives. The diagnosis of breast cancer often comes as a shock. The few days or weeks between receiving the diagnosis and starting treatment are filled are often filled with anxiety, confusion and a sense of a loss of control. Breast cancer is profoundly disruptive to women's busy lives, as they must quickly assimilate the news, cope with conflicting emotions, learn more about the disease, and shuttle between different appointments. This emotionally laden organizational context impacted the project's development. We needed to be mindful of the cost (both time and energy) imposed on patients – many of whom were already at their limits. We wanted to minimize the time patients needed to fill out surveys, avoid overloading them with interviews, and develop interventions that were sensitive to their needs.

We began this project with a qualitative analysis of how patients and physicians interacted in decision-making consultations. How did treatment decisions get made? What happened when things went well? What happened when things went poorly? How did patients and physicians communicate, and what prevented them from doing it well? Then we needed to develop an understanding of their goals or ideals for collaborative decision making, and the barriers to achieving the goals. What would be better according to patients? What would be better according to physicians? What prevented them from achieving their goals? What supported them in achieving their goals? What kinds of tools or services would help reinforce the support or overcome those barriers? What kind of constraints (resources, time, organizational) did we need to take into account? Finally, as we tried new interventions, we needed help understanding their impact on the current situation and whether or not we had moved closer to the goals of improving patient-physician communication and collaborative decision making.

Getting Started: The Social World of Patients and Physicians

To better understand the challenges patients and physicians faced in making treatment decisions, Karen Sepucha immersed herself in the clinic. She followed three doctors around for days, and interviewed them individually about their activities. She also interviewed several nurses and administrative staff, attended tumor board meetings and other physician conferences. In addition, Karen offered Consultation Planning services to patients, developed through an earlier action research project at Stanford University, the BCC and the Community Breast Health Project. Consultation Planning involves an hour-long discussion with patients prior to their appointment with a physician, in which we map out their questions, concerns, and issues into a flowchart (Sepucha, Belkora et al.; Belkora, 1997). Then the patient and physician use the Consultation Plan in the medical consultation to clarify questions, discuss all the issues at stake, and come to a decision about treatment.

Throughout this process, we analyzed the interdependencies and relationships in the clinic and explored the following questions: who are the participants? Who are the stakeholders? What decisions are they making? What are the authority relationships? How are resources shared or regulated? What are the channels of communication? What actually happens in those interactions where decisions are made? Mapping these practices was difficult, interpreting and determining what to do was even harder. The clinic setting required an investigation into micro-processes of how patients and physicians negotiated meaning, as well as more a macro approach to understand how broader social and economic changes affected the delivery of care (Clark, 1991).

Through our ethnographic observation, we began to identify and develop the concept of a "decision gap." Physicians had detailed protocols for diagnosing breast cancer, and detailed protocols for treating breast cancer, however they did not have any tools or structure for the period of time between diagnosis and treatment. Patients receive emotionally overwhelming news of a diagnosis and lots of clinical information, but no structured way to sort through their feelings, concerns, and priorities about treatment. During this time patients and physicians make critical decisions that will impact patients' bodies and lives.

The main interaction between patients and physicians during the decision gap occurs during the medical consultation. We began to document these critical consultations more thoroughly and to identify the barriers to productive collaboration. The following case is interpreted from a transcribed interaction between a patient and highlights several problems we observed. [A more in-depth discussion can be found in Sepucha, 1999.]

A patient, Ms. Murphy, is getting a second opinion from an oncologist. She had surgery two weeks ago and now needs to make a decision about adjuvant therapy, but everything she has heard seems to suggest that the decision will depend on her lymph nodes. Thirty minutes after her appointment was scheduled to start, Ms. Murphy wonders if anyone knows that she is waiting. Alone in the room, the knot in Ms. Murphy's stomach keeps growing. She is anxious about what the doctor will say. She is hoping for good news from the pathology report but fears that her cancer has spread.

At 3:25 p.m. Dr. Smith finishes seeing an unscheduled patient and quickly answers two pages. He glances at his schedule to see who is next, a new patient, Ms. Murphy, and then six more patients before 5 p.m. He glances at his watch and begins to flip through her chart and take furious notes. Her history is long and very complex, and to make matters worse, the copies of her records are not in order. With the hope that she can help him piece the history together, Dr. Smith grabs the chart and walks in to greet Ms. Murphy.

After a quick greeting, Dr. Smith immediately starts to ask many specific questions. During this time he barely looks up and instead is flipping back and forth through pages of her medical chart, asking for specific dates and things that have happened, taking notes.

Ms. Murphy wonders why she had to bring her things an hour early if he didn't even read them beforehand. She was also anxious to see if the doctor had her pathology report, so she can learn the status of her lymph nodes. Then she began to worry that he was stalling because it's bad news. After about 15 minutes, Dr. Smith looks up and starts to summarize what he just pieced together of her history. When he pauses, Ms Murphy finally asks a question,

Ms. Murphy: So what about the sentinel node pathology report?

Dr. Smith: All six of the other nodes were negative by microscopy, by looking at them. The special stain on those nodes is pending.

Ms. Murphy: That was supposed to be done today. It was supposed to have been done last Monday and the surgeon called yesterday and told me about the one you just mentioned and the other six were supposed to have been done today. Has anyone called about the pathology report?

Dr. Smith: I don't know. I just spoke to pathology now and we can call them again after we speak.

Ms. Murphy: Yes, that's the whole thing. We need that before we can talk.

Dr. Smith: Well to some extent, yes, it would certainly help us. But I can already tell you, the fact that the sentinel node is involved, it is a little hard to interpret because it is a new test, but nevertheless, it is positive. Currently our best estimate to assess the risk associated with that is to assume that it is the same as if it was positive by any other means.

Ms. Murphy: Now the surgeon spoke very differently about that, so I'm confused. The surgeon said it made no difference, the surgeon said that since there was no gross tumor, that it was as if it was negative.

Dr. Smith: Well that's why I want to stress that we don't know. This whole sentinel node procedure is new. That's when they get the node with the radioactive material and then finding the cancer, with the old technique, which is looking under the microscope or whether finding the cancer with this new technique, this special staining, we really don't know how this fits in.

Ms. Murphy: This is *really* anxiety provoking for me. The surgeon talks about it one way, you talk about it another way there is no definitive information for me here. Does anybody read my chart? Does anybody follow up on these phone calls? Does anybody care?

This scenario highlights several problems we consistently found during medical consultations. First, medical consultations lack some basic standards for a meeting – patients often do not know when consultations will start, or how long they will last. Lengthy waiting times and the uncertainty can create additional anxiety and frustration for patients coping with a recent diagnosis.

Physicians' overbooked schedules leave them barely enough time to read a patient's name on the chart before entering the room, and medical charts are disorganized and often missing important reports. Current standards for "the medical interview" focus on the clinical aspects of the case (Lipkin, Putnam et al., 1995). Medical charts reinforce this by containing only documentation of clinical information intended for other clinicians (e.g. as opposed to a list of patients' questions or descriptions of situations in a language patients can understand). Research suggests that by increasing the time spent reviewing the chart, Dr. Smith risks compromising Ms. Murphy's satisfaction and understanding (Smith, Polis et al., 1981).

Physicians are not the only ones who do not adequately prepare for the consultation. Patients often do not know what questions to ask. Even for patients who know what questions to ask, shyness, defensiveness and confusion can prevent them from interrupting the doctor (Sepucha, Belkora et al.; Roter, 1977; Roter, 1984) The lack of preparation prevents patients and physicians from voicing their questions and concerns. As a result, it is difficult to clear up confusion and decide on the best course of action. Not surprisingly, physicians did most of the talking. Doctors spoke rapidly and moved quickly from one issue to another, often slipping into medical jargon. Neither the patients nor the physicians asked many questions, and if they did, they were often closed questions that did not lead to further elaboration.

Patients did not leave the consultation with a record of the conversation that had been reviewed by the physicians for completeness or accuracy. Occasionally the patients (or a support person accompanying the patient) took notes. It was often unclear what had been decided and who was supposed to do what next. Patients often scheduled follow-up appointments or called the physicians to clarify issues previously discussed. In some cases things fell through the cracks (e.g. tests never got ordered, patients did not get enrolled in clinical trials). Physicians often complained that they needed to repeat the same information over and over to their patients.

From our exploration into the decision gap, three recurring themes emerged. Confusion, poor communication and overload consistently posed barriers for patients and physicians as they tried to make treatment decisions. Figure 1 synthesizes the specific problems and breakdowns that contribute to each of these themes. For example, patients often got confused by having to choose among undesirable alternatives and conflicting recommendations from physicians. Similarly, physicians often got overloaded trying to fill the multiple roles required during the consultation, (e.g. recorder, educator, facilitator, supporter, and expert). Confusion, poor communication and overload are dependent, e.g. a patient who is experiencing overload, is less likely to be able to voice her questions and concerns and communicate productively with her physician.

We shared our emerging analysis with patients and physicians, and reviewed previous consultation plans to see if the issue raised by patients could be linked to confusion, poor communication and overload. Many physicians would read through the summary descriptions and transcripts, and recognized different aspects of the situation. But most still thought the data described other physicians and other patients, not their own interactions. Documenting these dynamics

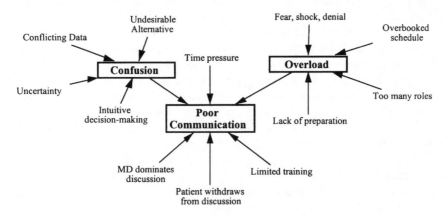

Fig. 1. Three Main Barriers to Collaborative Decision Making in Breast Cancer Consultations and the Factiors that Contribute to these Barriers. [Reproduced with Permission from Sepucha, 1999].

through audio-tapes and other means was critical, because, with a transcript, they could no longer tell themselves that it was someone else. As we analyzed their situation, they began to see things from a new perspective and became more receptive to the idea of changing their practices.

Consultation Recording Methods

Based on our qualitative analysis, our goal became to create a systematic method to make decisions and promote open, balanced communication between patients and physicians – and to do this without overwhelming or overloading patients and physicians. We wanted to create a process that bridged the gap between the knowledge physicians bring and the knowledge patients bring, in order to foster better interaction and help them create high quality, collaborative decisions during medical consultations.

Even with this focus, it was not immediately clear how to do that. Another breakthrough came when we decided to view the medical consultation as an important meeting. Neither patients nor physicians generally approach medical consultations as meetings, as neither usually schedules them, creates an agenda, or agrees together who is in charge. Medical consultations suffer from many of the unproductive dynamics commonly found in poorly run meetings, such as wheel spinning, layering of topics, defensiveness, and role conflicts (Doyle, 1982). We found that during consultations physicians tended to take control of the discussion, repeat the same information, layer topics, interrupt patients, and dominate. Overwhelmed and scared patients withheld their questions and concerns and withdrew from the discussion during consultations.

We found a simple and promising framework in Doyle and Straus' *How to Make Meetings Work* (1993). Doyle and Straus stress the need for four roles in every meeting. The *facilitator* is responsible for preparation and helps the group work together by making process suggestions (as opposed to content suggestions). The *recorder* publicly documents the ideas and questions that are voiced; the *group leader,* voices his or her concerns and questions and keeps the group focused on the agenda; and the *group members* voice their concerns and questions in a productive manner.

When we envisioned the medical consultation as a meeting with a facilitator and recorder, we saw the potential to improve consultations by reducing overload. A facilitator would help patients prepare for consultations (using Consultation Planning) and would make sure that patient's questions and concerns were addressed adequately. A recorder would free patients (and support them), enabling more attention for listening, understanding, and asking questions. The patient's preparation would help the physician better prepare for the consultation (e.g. if the patient indicates that the results of her pathology report is her number one question, the physician can make sure to track it down before going in). A record of the consultation in language patients can understand would (hopefully) reduce follow-up – and insure that patients leave with an accurate summary of biomedical information and next steps in their care. However, as described by Doyle and Straus, the facilitator/recorder role did not provide much explicit help overcoming barriers to communication or making complex decisions. As a result, we used methods and tools from action science, which promotes open communication (Argyris, 1993; Action, 1996), and decision analysis, which provides structure and methods to evaluate complex decisions (Howard & Matheson, 1989; Howard & Matheson, 1989; Keeney, 1992).

Consultation Recording methods introduced a facilitator/recorder into the consultation to help patients and physicians set an agenda, promote open communication, and create a written record of the consultation for all participants. Consultation Recording (CR) methods follow a five-step process. The flow of a consultation with CR is depicted in Figure 2. The medical chart and the patient's Consultation Plan are inputs to the consultation. Then patients and physicians progress through the five steps of Consultation Recording. Then patients and physicians receive a Consultation Record and the physician dictates a note for the medical chart.

The input to, process and output of the medical consultation took the following form:

First, before the consultation, a trained facilitator helps patients prepare during a Consultation Planning session. (See Belkora, 1997 for more details on Consultation Planning.) The Consultation plan includes details about the *process* patients want to follow to make decisions (e.g. "I want the doctor to make the

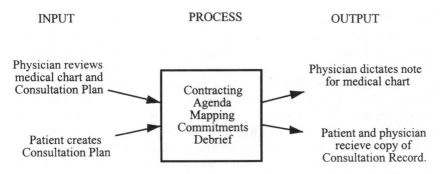

INPUT   PROCESS   OUTPUT

Physician reviews medical chart and Consultation Plan → Contracting Agenda Mapping Commitments Debrief → Physician dictates note for medical chart

Patient creates Consultation Plan → → Patient and physician recieve copy of Consultation Record.

Fig. 2. The Flow of a Medical Consultation with Consultation Recording.
[Reproduced with Permission from Sepucha, 1999].

decision but strongly consider my opinion.). It also articulates statements that detail the *content* patients want addressed such as, "I don't understand my pathology report, do I have invasive cancer or non-invasive cancer?" During the consultation, the patient, physician and a facilitator proceed through the five phases of Consultation Recording. In the **Contracting** phase, the group agrees on a process and clarifies the roles (including the patient's preference for participation in decisions). In the **Setting the Agenda** phase, the patient and physician agree on a common focus on the content for the consultation. In **Mapping**, the patients and physicians discuss the agenda items. The facilitator intervenes to maintain an open and balanced conversation and records the discussion using flowchart software. For **Commitments**, the facilitator helps the patient and physician explicitly document the decisions that have been made and any next steps, to be done, when, and by whom. During the **Debriefing** phase, the facilitator prints out the Consultation Record, which participants review to make sure they have covered everything on the agenda and in the Consultation Plan. This debriefing helps to close to the meeting, allows the physician to carefully edit any medical information in the Consultation Record, and helps reinforce the commitments and decisions of the group.

The Consultation Record summarizes the discussion. The record structures the salient issues, captures the information relevant to the decision, including choices and probabilities, and connects the reasoning. The records also documents the decisions or commitments that have been made and the next steps that need to be taken. The record is not a lengthy transcription of the conversation, rather it is a one page summary that structures the key information necessary to make and implement treatment decisions.

Experiences Working with Patients

Before this project, we had approached improving collaborative decision making by helping patients prepare for medical consultations. We helped patients think through their questions and concerns and printed out flow charts, or Consultation Plans, for patients to share with their physicians. Throughout this experience several things became clear. First, patients need help organizing and prioritizing their concerns. Second, information alone is not enough to enable patients to communicate with their doctor or make decisions. Finally the interaction nuances and quality of doctor-patient relationships are important (Sepucha, Belkora et al.) in making and implementing decisions (Goffman, 1967, 1981). However, until we started observing actual consultations of patients who had generated a Consultation Plan, we had little to no data on the impact in the consultation itself. After observing the medical consultation, we realized that preparing patients might not be enough to improve the quality of the collaboration. Although Consultation Recording methods sounded promising, we didn't know how an extra person in the room would impact the consultation, and whether a facilitator would improve or inhibit the consultation.

Karen's first case was difficult emotionally, socially, and medically. It highlights how the context and constraints of patients' lives influence their decision-making process. The patient was squeezed in at the end of a long day. Her appointment started around 6:30 and lasted until a little after 8pm. She was currently living in another country, and unsatisfied with doctors there, who told her she was going to die if she didn't have surgery immediately. She was a single parent, with two young kids who would need to be relocated if she stayed in the U.S. for treatment. Karen mostly recorded, and felt slightly overwhelmed by the intensity of the conversation. Half of the discussion focused on her lack of support system, her relationship with her ex-husband, job issues and financial concerns surrounding the location of her treatment. The other half dealt with "medical" issues including describing her medical situation, choices for treatment, and what the risks and benefits of the choices were. Her cancer was advanced and she needed to decide whether she wanted to start treatment with chemotherapy or surgery. Although most breast cancer patients have several weeks to think through their decisions, she needed to start treatment soon. The best choice depended heavily on taking financial and social issues into account.

During the discussion Karen struggled to record all these issues (and connect them to her decision about treatment – which the patient indicated as her main goal for the consultation in her Consultation Plan). Karen only participated in the discussion to check that the medical information she was transcribing was an accurate reflection of the doctors' comments, and did not interrupt to inquire into the patient's understanding. Nor did she ask how certain topics might or might not have been relevant to the decision at hand. Eventually Karen improved at condensing the conversations and actively facilitating the discussions. The concerns about the presence of another person in the room lessened. And the physicians even started adopting some of the methods into their style, by reviewing the CP and asking to set an agenda, and asking for their to-do list before leaving the room.

Negotiating Roles

The following discussion explores how emotions, interactions, and relationships were managed in this organizational context. Our experience provides suggestions and clues as to how researchers balance and negotiate multiple roles in a complex setting, how the interaction between researchers and partners changes over time, and how those changes impact the project's development.

Outsider or Insider

Karen's role at the BCC and relationships with the physicians and staff changed dramatically over the course of the project. She is not sure which came first – the change in her internal commitment or the change in the time she spent there. Initially, Karen felt the "distance" that is sometimes advocated in many researcher-subject relationships. With only a few hours spent at the clinic every week, it was hard to get a good sense for the system and really build a collaborative feel.

Karen decided to become more of a regular presence, which changed the dynamics. Doctors would pull her into consultations, or ask her to work with their patients. She observed many informal "backstage" conversations, and watched physicians talking in the hallway or the conference room – often right after a disconcerting situation had occurred. As time passed, she felt more invested in process, and connected to the people. Over the course of a few months, she accepted more ownership of the project, and felt a sense of responsibility for the quality of care delivered by the physicians and received by the patients.

Patient's Advocate or Physician's Assistant

Before coming to UCSF, Karen had completed some research projects that we developed in collaboration with a local community resource center. The center was created and staffed mainly by breast cancer survivors, who articulated a powerful critique of the medical establishment, and advocated self-help and patient empowerment. The women at the center repeatedly told horror stories of misdiagnoses, of poor interactions, and of manipulative relationships in their experiences navigating breast cancer treatment. The BCC showed us another version of the same story. Medicine is a "greedy institution (Coser, 1974) that demands enormous commitments of time and energy of its practitioners (Cassell, 1998). Physicians had no time for lunch, or even the restroom. We watched them decipher pages of cryptic handwritten notes and lab reports, frantically trying to find missing reports. We saw scheduling mistakes and overbooking, which demanded the impossible task of being in two places simultaneously. The physicians were busy, rushed, and constantly worried about time. They came to work early, stayed late, and worked on weekends to return pages, answer e-mails and attend to their patients' needs.

In this frenetic, complicated place, we also saw emotionally intense moments that occasionally happened when patients and their health care providers truly connected in a time of need (Scheff, 1990). A hug or a "doctor's prescription" for a nice dinner or massage could change the relationship. In Karen's volunteer work at the community center, she suddenly found herself sympathizing with doctors and trying to offer some explanations for some of the seemingly horrible behavior. This shift created ambivalence and internal conflict. Karen felt that her role was to help patients navigate this difficult time, make sure they were prepared and that the physicians answered their questions and concerns. On the other hand, she also felt that her role was to support the physicians so they could develop their abilities to connect with their patients. Often during the consultation, these goals came into conflict. Consultation Recording methods and the reliance on a "neutral" facilitator role did not adequately address this tension. And although the methods did espouse to support both patients and physicians, in practice the facilitation and recording were primarily directed at supporting the patients by making sure they got what they needed from the consultation.

From Research Project to Program Implementation

During Karen's dissertation defense, a professor challenged that the results from the experiment were due to her unique skills and capabilities, something he called "the Karen Effect." We disagreed, believing that the necessary skills to facilitate and record consultations can be taught and learned by others. In 1999, the BCC director decided to create a full-time program to implement Consultation Planning and Consultation Recording as a free service for all patients and physicians at the clinic. The following discussion explores how the implementation process unfolded and some of the unanticipated challenges of we encountered to create organizational change.

As an applied qualitative sociologist with experience in community-based and ethnographic research, Caryn Aviv came to the Breast Care Center in September 1999. After a month of observation, she began to work with patients and physicians on a daily basis. She also kept an ethnographic field journal to record interactions, and the emotional work (Hochschild, 1983; Scheff, 1990) she experienced with patients, family members, and physicians. Adding a sociologist brought a new perspective and different questions to the project.

In contrast to engineering and decision analysis, sociology begins with different assumptions about the social world (Collins, 1992), often uses inductive methods to conduct research (Glaser & Straus, 1967), and proposes alternative ways to create and evaluate change (Gottfried, 1996). Engineers, using deductive, quantitative research methods, develop practical and efficient improvements to existing systems. Decision analysts strive for clarity and consensus, tend to view decisions as individualistic, rational choices, and often consciously ignore social and historical contexts that drive decision making.

Qualitative sociological research often aims for "thick description," rich details, and analysis of natural settings, actors, relationships, and processes where power and knowledge are exercised unequally (Smith, 1989). Academic sociologists working in this tradition advocate recommendations for structural changes in society, but they often fall short on proposals for practical implementation. Sometimes the goal of qualitative research involves suggesting social change (DeVault, 1999; Stacey & Thorne, 1985), and sometimes it merely aspires to describe current conditions to improve our understanding of social settings. Applied sociologists take the insights of this discipline, and attempt to create small-scale change within organizations, often using participatory research methods. The addition of a sociological perspective required a learning curve for all participants, to create a shared understanding of the conditions of the clinic, a common conceptual vocabulary, and agreement on how to move forward with the program.

Despite these disciplinary differences, we saw the traditions of action research and applied sociology as complementary and synergistic. As Caryn observed medical consultations and slowly began to provide collaborative care services, the challenge of transition from research to clinical practice raised several issues. How can we pass along the skills for an emerging, innovative method that combines multiple disciplines? How can we integrate a pilot research project into the everyday clinic workflow with busy physicians and anxious patients? We focus on two issues in the implementation process, which have identified new areas to explore as we adapt these methods and improve our decision support tools. One area involves passing along important skills required for effective Consultation Recording. The second concerns the ambiguity and role negotiation of the Consultation Recorder.

Learning Consultation Recording and Facilitation Skills

Initially, Caryn struggled with the practical aspects of Consultation Recording. How can one write the details about a patient's case that physicians thought was most important (which varies from physician to physician), in a way that provides value to the patient? How does one determine what that value might be for patients, whose goals and needs differed from physicians? As she watched doctors and patients interact, Caryn likened Consultation Recording to spear fishing. During a consultation, the flow of information can rush by quickly, like a swiftly moving stream. Consultation Recorders have to develop sufficient skills to capture succinct chunks of the most vital information, statistics, patient preferences, and commitments – in other words, to carefully watch the stream and to throw their spears effectively to catch the right fish. Simultaneously, Consultation Recorders keep the discussion on track, check for patient understanding, help patients and physicians articulate decisions and outline concrete next steps. Juggling all these communication and writing tasks proved difficult and challenging.

With her background in symbolic interaction and ethnography, Caryn had learned to provide as much detail, nuance, and "thick description" (Geertz, 1974, Lofland & Lofland, 1995) as possible to understand a social setting. With thick description as her framework, she did not know what to write on Consultation Records that would meet the diverse needs of patients and physicians, and wondered how she could fit everything on one page. Her stated goal was to structure and record decisions, and capture elements of conversations. But she also wanted to richly describe and document interactions, which created unwieldy consultation records and differed from the recording focus. In contrast, the physicians wanted a clean, concise, and succinct Consultation Record that mapped out decision trees. Caryn's early Consultation Records suffered from "too many spears," catching all sorts of fish.

In the early consultation records, the pages were so cluttered it was difficult to know what to focus on. Additionally, the decision patients and physicians made about treatment was assumed, but not explicitly articulated. Sometimes physicians wanted clarity, brevity, and a focus on the presentation of statistical risks, while patients wanted comprehensive explanation and narrative to understand their choices. After trial and error, we realized Consultation Recorders need to balance and reflect the needs of both patients and physicians. They have to learn what important information needs to be on the record, and how much (or how little) description to include. They also need training and practice to learn the "filtering" skills to identify and address those different needs. One year later, Consultation Records look quite different (Fig. 3).

The changes in Consultation Records over time suggest two insights. First, these skills can successfully be passed on. Caryn learned to anticipate standard information physicians give about diagnosis descriptions, surgery and chemotherapy side effects, and recovery. Filling in these "data fields" quickly gives her more time to facilitate and summarize the discussion between patients and physicians about treatment options and preferences. Second, she learned that trying to create a record of thick description hampers her ability to actively facilitate the conversation, and is a more appropriate strategy for a participant observer whose sole focus is to conduct fieldwork. Caryn's Consultation Records reflected the learning curve in filtering, but also in negotiating her role as a facilitator/recorder and ethnographer in the applied setting.

Negotiating Roles

The Consultation Recording model stipulates a neutral role in meeting facilitation (viewing both patients and physicians as participants who need help). Caryn initially found the neutrality stance perplexing and uncomfortable, given

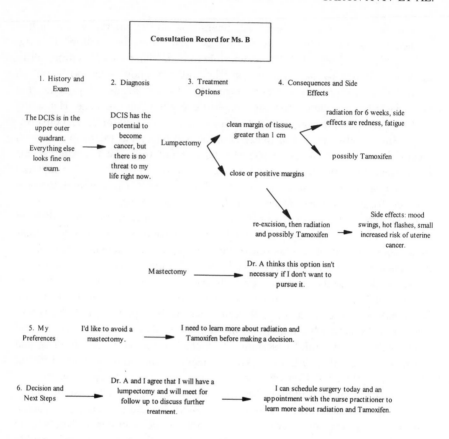

Fig. 3. A Consultation Record after One Year of Programme Implementation.

her training in medical sociology. That literature extensively documents physician dominance in conversations (Waitzkin, 1997), the reliance on professional distancing strategies of detachment, objectivity, and cultivated neutrality (Foucault, 1994; Lupton, 1994) and the persistence of inequalities based on age, race, economic class, sexual orientation, and gender (Lorber, 1997; Balshem, 1993). She questioned the assumption of neutrality, because power relationships between patients and physicians are so inherently unequal, especially when patients are women (West, 1993). Given this inequality, shouldn't the Consultation Recorder encourage more patient participation? Shouldn't the Consultation Recorder encourage physicians to listen more, and dominate less?

In addition to these theoretical and ideological questions about negotiating roles, there were pragmatic tasks to address. Karen Sepucha moved across the country. Caryn Aviv was on her own to integrate Consultation Recording and expand the number of physicians who used it, while Karen provided advice and support off-site. How did Caryn want to play the role of the Consultation Recorder, particularly with physicians who had never used the methods? Was she a patient advocate, neutral observer, or physician assistant?

Patients and physicians sometimes seemed to want different things, and Caryn brought more of a patient advocacy model to her work (Aviv, 1997). Some physicians asked her to perform administrative tasks that facilitated their work (and increased their appreciation for having someone else in the room), but required her absence from the consultation room during the discussion. Caryn worried that physicians occasionally enlisting her as their assistant blurred the role and detracted from her facilitation and recording responsibilities. Some patients wanted her to attend all their consultations as a patient advocate even after they had made decisions about treatment. Would working with patients throughout their treatment process dilute the original intent of Consultation Planning and Recording? These questions about role clarification and potential opportunities for more tools and support are areas we will explore in future research as we refine collaborative care methods and train others.

The implementation lessons we learned are twofold. Individuals bring different ideologies, agendas, experiences, and theoretical prisms through which to understand the dilemmas of collaborative decision making. An additional disciplinary perspective added complexity to the program, but also provided another window through which to critically view the context of decision making and patient-physician interaction. Second, negotiating a new role, program, and staff member into a busy clinic and the traditional dyad of patient-physician relationships is by no means easy. With a role model (Karen Sepucha) mentoring long distance, Caryn Aviv learned by doing, made mistakes, and solicited feedback. Combining applied sociology and Consultation Recording presented a struggle to blend divergent roles, theoretical frames, and practical approaches. We learned that in order to move from novice to skilled recorder/facilitator, implementation requires building agreement about roles, extensive observation of current conditions, and systematic training in theory, methods, and skills.

Collaborative Care Provides Support for Patients

We currently collect quantitative data from patient surveys that are completed after Consultation Recording, which assess satisfaction with consultation planning and satisfaction with the physician consultation. We have also conducted several focus groups to collect qualitative data and suggestions for program improvement. Our preliminary data suggest that patients strongly appreciate and benefit from the Program for Collaborative Care. Patients have written that they felt relieved another person was in the room to step in when they were feeling lost, overwhelmed, or scared. Others have provided detailed impressions of their experience and suggestions for bringing the program to other breast cancer clinics.

Two patients commented,

> "I think the most helpful aspect was what's called collaborative care. There's a woman who, before my appointment, she sat with me and asked me a lot of questions [about] what I knew and elicited the information that I had already found out, and then did a flow chart. We went in to see the doctor and she was there and she took notes on her laptop. She reminded me if there were questions that I had forgotten because you do. You know, all of a sudden you're hearing this news from the doctor and you forget what to ask. And then she had a flow chart at the end of all the information that I had been given, so I think that was the most helpful part." Allison S.

> "When I had the recurrence, I went to a medical center for a second opinion. I discovered that there is a space in a medical community where a group of people developed a system to support a patient and to listen to a patient and to treat a patient as an individual. That was so surprising, so astonishing to me that in fact it made me stop and realize that there was an absolute other way of being treated in the medical community. That in fact I could get the expertise and the compassion that I needed to take me through whatever choices that I made." Karen H.

Our data suggest that the Program for Collaborative Care improves the decision-making process for patients by providing timely support and structure to their medical consultations. We plan to create more comprehensive methods to evaluate the Program and assess how we can better meet the needs of patients and physicians, through survey revision, additional focus groups, and other assessment methods.

Collaborative Care as a Disruptive Innovation for Physicians

The physicians' response to the Program for Collaborative Care varied. Currently, two surgeons, four medical oncologists, and one radiation oncologist have used the program in their medical consultations. Reactions to the program have ranged from overt skepticism to indifference to enthusiastic support. We have learned that building consensus about the practical implementation of a new program and its potential benefit is critical to organizational change.

We conducted individual interviews with each physician in May 2000, to assess the implementation process and solicit direct, critical feedback. From the physician perspective, Consultation Planning and Recording methods raises issues of infringing upon professional autonomy, the effectiveness of physicians' interaction skills, and the question of disruptive intrusion into the medical consultation. As Caryn struggled with the Consultation Recorder's role, so did physicians. Some physicians wanted Consultation Recorders to step in frequently to clarify, keep conversations on track, and summarize, while others preferred an observer/recorder who did not interrupt the flow of conversation. Some physicians seemed defensive or irritated when she accompanied patients into the consultation. One physician said that while Consultation Plans were useful for anticipating patients' questions and concerns, Consultation Recording implied a deficit in her own communication skills. Having a third party, she explained, undermined her authority as a physician and facilitator, and she didn't think she needed it. Consequently, Caryn adapted the level of active facilitation to address physician needs, and in some cases, abandoned facilitation altogether to promote more physician support for the program. While this strategy lessened conflict, she sometimes questioned whether her work actively promoted more collaborative decision making between physicians and patients. Consultation Recording requires establishing cooperative relationships with physicians to encourage their participation, while paradoxically challenging their unquestioned authority. To facilitate a productive meeting, Consultation Recorders often need to interrupt, clarify, and reiterate complex information, which can raise the hackles of physicians who might have different ideas about the definition of non-intrusive facilitation. If the Program for Collaborative Care becomes implemented elsewhere, Consultation Recorders need to develop relationships with the physicians, to explain the program, observe diverse interaction styles, and solicit suggestions for how to best meet their particular needs prior to stepping into the Consultation Recording role. They will need to accommodate to individual physicians' needs and interaction styles to work effectively and collaboratively. Variation in physician interaction style requires flexibility, adaptation, and a wider array of tools at their disposal to promote collaborative decisions with patients.

From Implementation Back to Research

After one year of implementation, we have identified several areas for future research to improve our tools and methods for critical medical decision-making consultations, including. These include comprehensive patient support, more physician preparation and logistical support, better tools to record consultations, and better ways to translate clinical evidence into usable patient information. Based on the feedback from physicians, we realize that they need more help to engage patients collaboratively. Consultation Recording is time-intensive, challenging to implement with physicians, and requires significant time and resources to train facilitators. One goal is to improve the quality of collaborative breast cancer treatment decision making by improving the decision support tools physicians use in their consultations. Future research will focus on how to transition collaborative care to non-facilitated tools and methods, in order to reduce cost, streamline the process, take advantage of new technologies, and promote easy replication.

We have also identified a gap in the decision and emotional support for women with metastatic breast cancer (which has spread to other parts of the body) and their physicians, who often face very different decisions about treatment and end of life issues (Mamo, 1999). The culture of medicine and patient's fears about their own mortality pose challenging barriers to openly discussing the risks and benefits of end stage treatment options and feelings about end-of-life issues (Larson, 2000; The, 2000). Future research will use ethnographic and action research methods to examine how oncologists and patients currently navigate this process, to assess how collaborative care, in conjunction with new emotional and psychological support interventions, might address the needs of patients, physicians, and families facing metastatic cancer and the dying process.

We have found action research to be a useful approach to the formulation, implementation and assessment of interventions in breast cancer decision making. Our partnership between researchers and clinicians has yielded fruitful interventions that try to help patients and physicians navigate the complexities of communication, information, and treatment options. Action research provided a framework to describe the challenges patients and physicians face in making treatment decisions. It also helped us develop a viable intervention (Consultation Recording) with a motivated client based on our analysis of patient and physician needs. Finally, we used this approach to implement the intervention into a busy clinic, and to evaluate the transition from research to a full-time program. We identified some key tasks that are necessary to move an action research project from research into practice, to insure successful implementation that sticks. We learned about core skills necessary to train future effective Consultation Recorders, how to incorporate physicians, and how to modify our methods to fit physician style. Action research also helped us identify new areas of need for better patient and physician support.

This case study makes two claims. First, an action research framework can successfully help researchers and their clients identify needs, develop and implement interventions, and assess their value in changing the status quo. Our experience suggests that this approach (while by no means easy) helped us address a significant, persistent problem in health care, changed how patients and physicians interact, and facilitated more collaborative decision making. Second, action research can create a feedback loop to further identify and address emerging needs in an applied health care setting. If they successfully manage the tensions between academia and organizational needs, sociologists, anthropologists, and other health researchers can pragmatically address important social problems. Action research can be used to expand research projects beyond their initial parameters, sustain research relationships, and affect the way patients and physicians negotiate care in rapidly changing, complex social settings.

ACKNOWLEDGMENTS

This research was supported by grants from Arthur Vining Davis Foundations, the National Science Foundation, and the Department of Defense.

REFERENCES

Action, D. (1996). Organizational Learning in Action. Training Materials.

Argyris, C. (1993). *Knowledge for Action: A Guide to Overcoming Barriers to Organizational Change.* San Francisco, Jossey-Bass.

Aviv, C. (1997). *Having You There Made It Better: The Feminist Emotion Work of Rape Crisis Counselors. Department of Sociology/Anthropology.* Chicago, IL, Loyola University Chicago.

Balshem, M. (1993). *Cancer in the Community: Class and Medical Authority.* Washington, D.C., Smithsonian Museum.

Belkora (1997). Mindful Collaboration: Prospect Mapping as an Action Research Approach to Planning for Medical Consultations. Engineering-Economic Systems and Operations Research. Stanford, Stanford University.

Cancian, F. (1996). Participatory Research and Alternative Strategies for Activist Sociology. In: H. Gottfried (Ed.), *Feminism and Social Change: Bridging Theory and Practice* (pp 187–205). Champaign, IL, University of Illinois Press.

Cassell, J. (1998). *The Woman in the Surgeon's Body.* Cambridge, MA, Harvard University Press.

Clark, J. A., Potter, D. A., & McKinlay, J. B. (1991). Bringing Social Structure Back into Clinical Decision Making. *Social Science and Medicine, 32*(8), 853–866.

Collins, R. (1992). *Sociological Insight: A Guide to Non-Obvious Sociology.* Oxford, Oxford University Press.

Coser, L. (1974). *Greedy Institutions: Patterns of Undivided Commitment.* Riverside, NJ, The Free Press.

Devault, M. L. (1999). *Liberating Method: Feminism and Social Research.* Philadelphia, Temple University Press.

Fisher, S. (1996). *In the Patient's Best Interest: Women and the Politics of Medical Decisions.* New Brunswick, NJ, Rutgers University Press.

Doyle, M. A. D. S. (1982). *How to Make Meetings Work.* New York, Jove Books.

Foucault, M. (1994). *The Birth of the Clinic: An Archaelogy of Medical Perception.* New York, Vintage Press.

Geertz, C. (1974). *The Interpretation of Culture.* New York, Basic Books.

Glaser B., & Straus, A. (1967). *Discovery of Grounded Theory: Strategies for Qualitative Research.* New York, Walter De Gruyter Press.

Goffman, E. (1967). *Interaction Ritual.* New York, Anchor Doubleday.

Goffman, E. (1981). *Forms of Talk.* Philadelphia, University of Pennsylvania Press.

Gottfried, H. (Ed.) (1996). *Feminism and Social Change: Bridging Theory and Practice.* Urbana, IL, University of Illinois Press.

Hart, E., & B. M. (1999). *Action Research for Health and Social Care: a guide to practice.* Philadelphia, Open University Press.

Hochschild, A. (1985). *The Managed Heart: The Commercialization of Feeling.* Berkeley, University of California Press.

Howard, R. A., & Matheson, J. E. (Eds), (1989). The Principles and Applications of Decision Analysis. Menlo Park, Strategic Decision Group.

Howard, R. A., & Matheson, J. E. (Eds), (1989). The Principles and Applications of Decision Analysis. Menlo Park, Strategic Decision Group.

Keeney, R. L. (1992). *Value-Focused Thinking: A Path to Creative Decisionmaking.* Cambridge, Harvard University Press.

Klawiter, M. (1999). Racing for the Cure, Walking Women, and Toxic Touring: Mapping Cultures of Action within the Bay Area Terrain of Breast Cancer. *Social Problems, 46*(1), 104–126.

Larson, D. (2000). Anticipatory Mourning: Challenges for Professional and Volunteer Caregivers. In: T. A. Rando (Ed.), *Clinical Dimensions of Anticipatory Mourning: Theory and Practice in Working with the Dying, Their Loved Ones, and Their Caregivers* (pp. 379–398). Champaign, IL, Research Press.

Lewin, K. (1951). *Field Theory in Social Science.* New York, Harper & Row.

Lipkin, M. J., Putnam, S. M. et al. (Eds) (1995). *The Medical Interview.* New York, Springer-Verlag.

Lofland, J. A. L. L. (1995). *Analyzing Social Settings: A Guide to Qualitative Observation and Analysis.* Belmont, CA, Wadsworth.

Lorber, J. (1997). *Gender and the Social Construction of Illness.* Thousand Oaks, CA, Sage Publications.

Lupton, D. (1994). *Medicine as Culture: Illness, Disease, and the Body in Western Societies.* Thousand Oaks, Sage.

Mamo, L. (1999). Death and Dying: Confluences of Emotion and Awareness. *Sociology of Health and Illness, 21*(1), 13–36.

McCormick, A., McKay, M. M., Marla, W., McKinney, L., Paikoff, R., Bell, C., Baptiste, D., Coleman, D., Gillming, G., Madison, S., & Scott, R. (2000). Involving Families in an Urban HIV Preventive Intervention: How Community Collaboration Addresses Barriers to Participation. *AIDS Education and Prevention, 12*(4), 299–307.

Nyden, P., Figert, A., Shibley, M., & Burrows, D. (1997). Building Community: Social Science in Action. Thousand Oaks, CA, Pine Forge Press.

Potts, L. (1999). *Publishing the Personal: Autobiographical Narratives of Breast Cancer and the Self. Ideologies of Breast Cancer.* L. Potts. New York, St. Martin's Press.

Roter, D. (1977). Patient participation in patient-provider interactions: effects of patient question asking on the quality of interactions, satisfaction, and compliance. *Health Educ. Monographs,* 280, 535.

Roter, D. (1984). Patient question asking in physician-patient interaction. *Health Psychol., 3*(5), 395–409.

Scheff, T. J. (1990). *Microsociology: Discourse, Emotion, and Social Structure*. Chicago, University of Chicago Press.

Sepucha, K., Belkora, J. et al. (submitted). Consultation Planning to Help Breast Cancer Patients Prepare for Medical Consultations: Effect on Communication and Satisfaction for Patients and Physicians.

Sepucha, K., Belkora, J. et al. (2000). Building Bridges Between Physicians and Patients: Results of a Pilot Study Examining New Tools for Collaborative Decision Making. *Journal of Clinincal Oncology, 18*(6), 1230–1228.

Sepucha, K. C. (1999). Consultation Recording Methods to Facilitate Collaborative Decision-Making in Breast Cancer. Engineering-Economic Systems and Operations Research. Stanford, Stanford University.

Smith, C., Polis, E. et al. (1981). Characteristics of the initial medical interview associated with patient satisfaction and understanding. *J. Fam. Pract., 12*(2), 283–288.

Smith, D. (1989). *The Everyday as Problematic: A Feminist Sociology*. Boston, Northeastern University Press.

Stacey, J., Thorne, B. (1985). The Missing Feminist Revolution in Sociology. *Social Problems, 32*(4), 301–316.

The, A., Hak, T., Koeter, G., & Van der Wal, G. (2000). Collusion in Doctor-Patient Communication about Imminent Death: An Ethnographic Study. *Behavioral Medicine Journal, 321*, 1376–1381.

Waitzkin, H. (1997). *The Politics of Medical Encounters: How Patients and Doctors Deal with Social Problems*. New Haven, Yale University Press.

West, C. (1993). Reconceptualizing Gender in Physician/Patient Relationships. *Social Science and Medicine, 36*(1), 57–66.

White, S. (Ed.) (1995). *The Cambridge Companion to Habermas*. New York, Cambridge University Press.

HEALTH DETERMINANTS AMONG PERSONS WITH TRAUMATIC INJURIES: A COMPARATIVE ANALYSIS USING THE SICKNESS IMPACT PROFILE

Jeffrey Hall and William C. Yoels

ABSTRACT

The purposes of this study were to: (a) measure and compare the subjective health status of persons with spinal cord injuries (SCIs); traumatic brain injuries (TBIs), burns, and intra-articular fractures of the lower extremities (IAFx); (b) identify factors that explain variation in the health assessments of persons with each injury type; and (c) determine whether the effects of study factors and variables are consistent across injury types. These tasks were accomplished by employing the Sickness Impact Profile (SIP), a generic, overall health status indicator, as the outcome measure in three separate sets of hierarchical regression analyses. The variable subsets useful in explaining overall SIP scores for the four injury types were defined and questions concerning similarities among the different injury types were addressed. Findings indicate that: (a) the overall, physical, and psychosocial health assessments of persons with TBIs, IAFx, and burns do not differ

Changing Consumers and Changing Technology in Health Care and Health Care Delivery,
Volume 19, pages 231–259.
2001 by Elsevier Science Ltd.
ISBN: 0-7623-0808-7

significantly; and (b) persons with SCIs, TBIs, IAFx, and burns should be considered distinct groupings with respect to the individual determinants of overall health status and the variable domains explaining the greatest amounts of variance in health status assessments.

INTRODUCTION

Within medical care, persons with different injuries are generally perceived and responded to as vastly different groups. One of the explicit aims of all sociological research is to expose the "general that can be found within the specific." That is to say, one of sociology's primary goals is to identify behaviors, processes, or relationships found across a variety of social contexts. Consistent with this objective, the central purpose of the present research was to discover the points of overlap in the experiences of persons with Spinal Cord Injuries (SCIs), Traumatic Brain Injuries (TBIs), Burns, and Intra-articular Fractures of the Lower Extremities (IAFx). Such injury groups are often thought to be as different as "apples and oranges."

There are numerous contexts of interaction where persons with the four injury types of interest in this study may be exposed to similar circumstances, encounters, or influences. One useful tactic for identifying these contexts involves analyzing the various points where there is interaction between the injured and the agents of assorted health care systems: during the acute (or secondary) and post-acute (tertiary and/or long-term) phases of care. This study examines the health statuses reported during the post-acute care phase and implications that dispositions held during the primary and acute care interaction phases may hold for functioning during this latter period. As life-spans continue increasing along with rapid improvements in medical and diagnostic technologies, it becomes important to understand how diverse groups of persons experience illness and injury. Such knowledge will improve the delivery of health care services to these populations.

One simple question that arises when one begins to think about how persons with injuries as seemingly different as SCIs, TBIs, Burns, and IAFx, may be compared is "Are the health status assessments of persons with SCIs, TBIs, IAFx, and burns similar or significantly different?" Knowledge here is severely limited, as much research in the area of injury focuses on either systemic factors (where individuals with different types of injuries are treated as aggregates) or on aspects only relevant to specific conditions. Moreover, the modest knowledge that does exist in this area includes either clinical conclusions (based on morbidity data) or assumptions based on anecdotal, common sense ideas about the severity of particular injuries. An alarming reality that becomes apparent after considering the previous statement is that such incomplete knowledge may make it difficult for these groups to access the therapeutic and rehabilitative services that might enable greater levels of functioning to be recovered.

A comparative analysis of the health states of four injury groups is the most appropriate means for obtaining an answer to the aforementioned research question. One approach to carrying out this task could involve a comparison of physician's assessments of the health of persons with SCIs, TBIs, IAFx, and burns. This approach is characterized by two very important drawbacks. First, physician-based assessments of health status are most likely to be primarily medically oriented, and thus permeated with statements focusing on physiologic health descriptors while ignoring the psychological and social statuses emphasized in the World Health Organization's conceptualization of health. Secondly, physician-based assessments cannot present the types of health characterizations created after persons subjectively deal with their health conditions.

In light of these problems, a second measure was used to capture and facilitate comparisons of health status assessments at two years after the acute-care discharge of persons with one of the aforementioned injuries: a socio-medical health index known as the Sickness Impact Profile (SIP) (Bergner, Bobbitt et al., 1976; Bergner, Bobbitt, Carter & Gilson, 1981; De Bruin, De Witte, Stevens & Diedriks, 1992; Elinson & Siegmann, 1979; Larson, 1991; McDowell & Newell, 1996). The SIP was selected to measure subjective health status because it provides a standardized means for: (a) collecting injured persons' thoughts concerning their performance in key areas that may be significantly impacted by injury, (b) for synthesizing these thoughts into global descriptions of functional ability, and (c) for quantitatively ranking the resulting descriptions. Our selection was also influenced by this instrument's standing as a psychometric "gold standard" (De Bruin et al., 1992; Larson, 1991; McDowell & Newell, 1996) and its underlying multi-dimensional, performance-based conception of health.

Two additional questions that may arise when considering the connections among persons with different injury types call attention to the topic of health status determinants: the influences responsible for variations in the outcomes of persons with SCIs, TBIs, IAFx, and burns. They include: "Given the few clinical similarities exhibited by SCI, TBI, IAFx, and burn patients, are there any shared variable domains that might impact the health states of these groups?" In addition, "How do these factors rank in terms of their contributions to explaining the health status of each injury group?" Literature searches using the following criteria were conducted as a first attempt to gain insight into these questions. Studies had to: (1) use the SIP as an outcome measure; (2) report information regarding both dimensional and total scores; and (3) be published between the 1981 and 2000. These searches revealed that although over 530 studies have used the SIP in various capacities, only a few investigations have used the SIP to measure the health status of persons with SCIs, TBIs, IAFx, or Burns. In total, only nine studies met our review inclusion criteria.

Of these nine, three studies focused on persons with burns (e.g. Patterson, Boltwood, Covey, De Lateur, Dutcher, Heimbach & Marvin, 1987; Ptacek, Patterson, Montgomery & Heimbach, 1995; Questad, Patterson, Heimbach, Ducher, De Lateur, Marvin & Covey, 1988), two examined the outcomes of spinal cord injury patients (e.g. Elliot, Herrick, Witty, Godshall & Spruell, 1992; Lundqvist, Siosteen, Bloomstrand, Lind & Sullivant, 1991) and one study centered on those having traumatic brain injuries (e.g. Moore, Stambrook, Gill & Lubusko, 1992). No studies were obtained for SIP-related searches focusing on intra-articular fractures. Consequently the three remaining studies focused on the outcomes of a surrogate population: persons with lower extremity fractures (LEFs); (e.g. Butcher, Mackenzie, Cushing, Jurkovich, Morris, Burgess, McAndrew & Swiontkowski, 1996; Jurkovich, Mock, Mackenzie, Burgess, Cushing, De Lateur, McAndrew, Morris & Swiontkowski, 1995; Mackenzie, Burgess, McAndrew, Swiontkowski, Cushing, De Lateur, Jurkovich & Morris, 1993).

Although the aforementioned studies generated important findings, they provide only limited information about the health status determinants that might be shared by persons with SCIs, TBIs, IAFx, and burns. Consequently, social research was consulted for potential answers. This body of research has provided major contributions to knowledge about the common determinants of health status across populations. These contributions provide directions concerning significant variables that are universally experienced by all persons interacting with health care systems (including persons with traumatic injuries).

For example, comparative studies of general population health suggest that group functioning is influenced in varying degrees by: (a) socio-demographic identities, (b) psychosocial dispositions, and (c) the level of access granted to activities and services aimed at health preservation and restoration. Despite the physiologic differences and diversities in professional responses to persons with SCIs, TBIs, IAFx, and burns, these groups are united by the fact that all are affected by these broad classes of variables (along with relevant clinical characteristics such as injury severity level and intentionality). The status sets formed by various combinations of clinical and socio-demographic attributes determine the levels and types of primary, acute, and long-term care provided to patients with each type of injury, and influence the amount of "social worth" implicitly assigned to persons in each group. Psychosocial dispositions determine whether injured persons must endure adaptive processes in relative isolation and rely exclusively on their own physical, psychological, and financial resources to facilitate health improvement, or whether they may benefit from the companionship, resources, and regulating influences of health professionals and significant others. Finally, differentials in health care access and exposure opportunities are indicative of each group's chances to benefit from restorative, skilled-care, and rehabilitative services.

THEORETICAL MODEL AND HYPOTHESES

Uniformity Among Health Status Determinants

Five variable domains were expected to influence the overall health status of persons with SCI, TBIs, IAFx, and burns: the (1) clinical, (2) access to care, (3) level of care, (4) socio-demographic, and (5) psychosocial domains. First, the variables comprising the clinical domain, such as intentionality and injury severity, were expected to be positively associated with overall SIP scores (i.e. to negatively affect participant outcomes). It was hypothesized that poorer health status would be associated with the possession of more severe injuries and intentional injuries. Second, insurance status and physician specialty, the variables constituting the access to care domain, were also expected to be positively associated with SIP score. We expected indigent participants and participants seen by a physical medicine and rehabilitation specialist to display worse health status than subjects covered by some form of insurance and subjects not examined by a rehabilitation physician. Third, higher SIP scores were expected for individuals exposed to higher levels of care. Specifically, we hypothesized, that participants discharged to formal care and referred to rehabilitation would report lower or worse health states than those released to home care and those not receiving rehabilitation.

Fourth, three variables in the socio-demographic domain were expected to be negative determinants of health status (age at onset, race, and alcohol involvement), while three others were expected to be positive determinants (education, gender, and marital status). We hypothesized, that whites, males, the married, the more educated, the employed, and participants with lower ages and no alcohol involvement in the injury producing event would display lower SIP scores than their polar opposites. Finally, the variables comprising the psychosocial domain, family satisfaction (a proxy measure of the adequacy of social support) and self-blame, were expected to share negative relationships with overall health status. It was hypothesized that higher levels of family satisfaction and blaming oneself for one's injuries would be associated with lower SIP scores, that is, better health outcomes.

Variable Domains: Assumptions about Relative Importance

One of this study's basic assumptions is that the clinical, access to care, level of care, socio-demographic, and psychosocial variable domains do not contribute equal amounts of information towards the explanation of overall health status. Here, it is assumed that these domains can be assigned priorities based on the specific conception of health providing guidance for one's research.

The biomedical and sociocultural conceptions of health constitute two frameworks that typically suggest two divergent paths of action. Within the current research, these conceptions are fused to suggest a course of action that allows the best insights of both perspectives to be used to construct propositions about the rankings among the five variable domains. We emphasize the ways in which health is affected by interactions between physiologic and social statuses. The clinical domain should emerge as the most important determinant of health because variables such as injury severity and intentionality influence more than just physical and psychological potentials for recovery; they affect the outcomes of provider-patient interactions in secondary care settings and have a substantial bearing on decisions regarding the likelihood that patients will benefit from tertiary forms of care (Caplan, Callahan & Haas, 1987; Haas, 1988; Wrigley et al., 1994). Clinical statuses are thus linked to both the physiologic and psychosocial difficulties that must be conquered at different points in recuperation and adaptation.

The access and level of care domains are expected to be the second and third most important health status determinants. In an ideal world, all patients would be able to take advantage of services that would place them in the best positions to reattain the functional levels exhibited prior to the onset of injury. However, the economic reality of medical care in the U.S. demands that access to certain services only be granted to those with the ability to pay for them, and patients' opportunities to benefit from the insights of physicians specializing in rehabilitation medicine are restricted due to the limited availability of individuals who have adopted this specialty. Injured persons who are not independently wealthy and simultaneously uninsured are placed at a considerable disadvantage in struggles for the limited resources found in the acute and post-acute contexts of care (Bamford, Grundy & Russell, 1984; Koska, 1988). Furthermore, persons who are unable to interact with rehabilitation physicians are less likely to receive essential rehabilitation services (Wrigley et al., 1994). Consequently, patients' statuses with respect to access variables may either complicate or simplify a recovery process already complicated by issues associated with clinical factors.

Variables reflecting the levels of care received by patients simultaneously indicate the extent to which functioning has been compromised, the degree of help needed to restore functional independence, and the nature of providers' judgments about the continued need for professional aid. Naturally, persons with more severe injuries are more likely to require higher levels of care and more extensive contacts with the health care system. But this observation becomes considerably more significant when the fact that some patients are not referred to services commensurate with their actual levels of need is considered. This mismatch may be attributed to the rigidity and insensitivity of the objective clinical criteria routinely used to determine whether injured persons should be granted access to certain treatment contexts. Though it is commonly assumed that persons discharged to lower levels of care possess the physiologic and psychosocial reserves necessary to recover without professional aid, such an assumption fails to account for the significance of services such as physical and occupational therapies for smoothing the transition back into daily living. Nevertheless, the functional recoveries of most persons referred to lower levels of care should still outpace those of persons referred to formal care facilities and to any form of rehabilitation. Ideally, differences in the level of difficulties faced in resuming normal activities should be manifested as discrepancies in patients' subjective assessments of health.

Finally, the sociodemographic and psychosocial domains should rank fourth and fifth among the five factors explaining health status variations. These low rankings are based upon the fact that variables constituting these domains do not directly impact health status. Instead they indirectly (and thus less strongly) affect health status by either increasing or decreasing the injured persons': (a) vulnerability to adverse events; (b) performance of health promoting behaviors; (c) contacts with hazardous physical and social environments; and (d) opportunities to access health enhancing activities and resources (Kaplan, 1989). Sociodemographic designations act as labels that interact with physical dispositions to structure the life chances of persons with injuries. Psychosocial statuses are embodiments of the psychological and social reserves that can be used to combat pathologies and buttress efforts to regain competence in performing the routine behaviors.

RESEARCH DESIGN AND METHODS

Study Sample

The study sample ($N = 463$) was drawn from a larger longitudinal study ($N = 3,156$) currently being conducted by the UAB-ICRC. Data were obtained from a retrospective acute care medical record review and annual follow-up telephone interviews with participants. These data were collected using a standardized instrument. The criteria for inclusion in the longitudinal study were: (a) having sustained one or more of the four injuries being studied; (b) having an acute care length of stay of 3 or more days; (c) residing and having been injured in Alabama; (d) being discharged alive from an acute care facility between October 1989 and March 1991; (e) being at least 18 years of age when injured; and (f) participating in scheduled, 12- and 24-month follow-up interviews. There were 1087 persons eligible to participate in the 24-month follow-up interview at the time the data were analyzed, 526 (48%) of whom actually participated in the SIP portion of the study. Forty-eight subjects were eliminated from the sample because 12-month data were missing from their questionnaires. Another 15 were excluded due to their possession of more than one traumatic injury (i.e. individuals with polytraumas). Consequently, the sample for this study includes information on 463 persons: SCIs ($n = 50$), TBIs ($n = 211$), burns ($n = 91$), and IAFx ($n = 111$).

The overall longitudinal study population was drawn from injured persons discharged from any one of nine hospitals representing a cross-section of patients in north-central Alabama. The participating hospitals included 5 of the state's 15 trauma centers (including one Level 1 Trauma Center) and 3 of 15 hospitals providing inpatient or outpatient rehabilitation services. Four hospitals were located in counties with large urban centers, which have historically served as referral centers for patients throughout the state. The remaining hospitals represented more rural areas located in the northern portion of the state and were the site of either trauma or rehabilitation services for their geographic catchment area.

Measurement

Dependent Variable

This analysis examined one dependent variable – overall health status – measured using the SIP (Bergner et al., 1981; Bergner, Bobbitt et al., 1976; Bergner, Pollard et al., 1976). The SIP was designed to serve as a "behavior-based instrument for measuring the impact of sickness that is comprehensive in scope and applicability, and that is sensitive in detailing the kind and degree of impact" (Bergner, Bobbitt et al., 1976: 399). It is based on the performance conception of health (Bauman, 1961), where subjects' health status assessments are expressed in terms of behavioral changes rather than changes in personal feelings or objective symptoms. For instance, statements such as "I walk shorter distances or stop to rest often," and "I am avoiding social visits from others" are respectively used to measure changes in physical and social performance.

The SIP version used in this study (Bergner et al., 1981), consists of 136 items that are classified into the following categories: (a) ambulation, (b) mobility, (c) body care and movement, (d) social interaction, (e) communication, (f) alertness, (g) emotional behavior, (h) sleep and rest, (i) eating, (j) work, (k) home and management, and (l) recreation and pastimes. Additionally, categories a–c can be combined to constitute the physical dimension, and categories d–g may be used to construct a psychosocial dimension. Lastly, scores may be calculated for each category, for the two dimensions, and for the full instrument. Dimensional and overall scores range from 0 to 100. Scores from 0 to 3 indicate no or little dysfunction, 4 to 9 designate mild dysfunction, 10 to 19 suggest moderate dysfunction, and scores 20 or higher represent severe dysfunction.

Independent Variables
Five sets of independent variables were examined. First, two clinical variables were measured: intentionality was measured as whether the injury was unintentional or the result of interpersonal or self-inflicted violence; injury severity was evaluated using the abbreviated injury severity scale. Second, two access variables were utilized: insurance status and physician specialty. Both variables were dichotomous: the first assessed whether or not a subject possessed medical insurance; the second measured whether or not the subject was seen by a physiatrist. Third, the level of care variables included discharge disposition, which was measured using a dichotomous variable indicating whether a subject was discharged to home (e.g. to self-, non-skilled, or home health care) or to a formal care facility (e.g. to a nursing facility or to formal inpatient rehabilitation); and rehabilitation, indicated by whether or not any form of rehabilitation, such as in patient, outpatient, or vocational, was received. Fourth, socio-demographic characteristics included age, alcohol involvement, education, employment status, gender, marital status, and race. Fifth, two psychosocial variables were measured: family satisfaction and self-blame. The Family Satisfaction Index (Olson, Sprenkle, & Russell, 1979) was used to assess this variable. It can be scored at several levels of aggregation providing scores for 14 items in two subscales and a total score. Scores for the individual items range from 1 to 5 and scores for the total instrument range from 14 to 70. Higher levels of family satisfaction indicate the presence of higher levels of social support. Self-blame was measured by whether or not participants blamed themselves for their injuries.

Statistical Analyses

Hierarchical multiple regression techniques were used to define the sets of variables useful in the explaining the overall SIP scores for each of the four injury types. The variable blocks, that is, the domains described in the previous section, were entered into four regression models in the following sequence: clinical variables, access variables, level of care variables, socio-demographic variables, and psychosocial variables. These steps were taken based on the assumption that the proportion of variance in SIP scores explained by the blocks entered in the earlier steps would be greater than that explained by blocks entered in the latter steps. It was our desire to compare the effects of clinical variables to the effects of all others. These comparisons were facilitated by assessing the amount of variation explained by each specific variable domain upon its point of entry into the regression models. We also wished to determine which independent variables were associated with SIP scores after controlling for other independent variables. Therefore, the individual relationships between the variables in the models and the outcome variable were evaluated by analyzing standardized and unstandardized regression coefficients, and t statistics.

RESULTS

Subjects

The characteristics of the four study samples are displayed in Table 1. Examination of this table shows the four sample populations to be approximately similar in terms of mean age, gender, and racial composition. The average age of subjects with SCIs, TBIs, and burns centered around 40 years, and subjects with IAFx tended to be slightly older, displaying a mean age closer to 50 years. With respect to gender and racial composition, most of the study's subjects were male and White, with lesser numbers of females and other ethnic groupings. The only sample where females constituted a clear majority (52% female, vs. 48% male) was that comprising individuals with IAFx. Finally, the spinal cord injury sample contained the highest percentage of non-White individuals. This figure is somewhat illusory, however, in view of the small size of this sample. In terms of relative numbers, the TBI sample contained the highest frequency of non-White subjects.

Whereas most of the subjects with SCIs, IAFx, and burns were married at 12 months post-discharge, over half (57%) of the subjects with TBIs were unmarried. Correspondingly, the last group contained the smallest percentage (43%) of married individuals, and the third (i.e. burns) contained the highest (62%). The SCI and IAFx samples were located at midpoints between these two extremes, consisting of 52% and 55% married, respectively. A large proportion of the study's subjects were unemployed at 12 months post-discharge. The burn sample was the only group where the employed constituted the

***Table 1*.** Characteristic of the Four Study Samples.

	SCIs ($n = 50$)	TBIs ($n = 211$)	IAFx ($n = 111$)	Burns ($n = 91$)
Mean age at onset	40 ± 15	40 ± 18	49 ± 18	40 ± 16
Age ≤ 25 years old	20%	28%	12%	20%
26 – 35	26%	22%	18%	21%
36 – 50	30%	25%	19%	34%
Age > 50	24%	27%	51%	25%
Race or ethnic group				
Non-White	26%	21%	22%	25%
Gender				
Male	74%	67%	48%	80%
Alcohol involvement				
Involved	26%	26%	10%	9%
No insurance				
Indigent	34%	23%	17%	18%
Discharge disposition				
Discharged to formal care	76%	26%	6%	13%
Intentionality				
Intentional	10%	10%	2%	4%
Marital status at 12 months				
Married	52%	43%	55%	62%
Employment status at 12 months				
Employed	14%	43%	35%	57%
Blame self for injury				
Yes	32%	21%	15%	63%
Education				
High school and above	50%	69%	73%	64%
Rehab. physician consulted				
Seen by a rehab. Physician	54%	8%	-	4%
Rehabilitation received				
Yes	86%	49%	64%	63%
Abbreviated injury scale	3.22 ± 1.68	2.94 ± 0.91	2.47 ± 0.52	2.78 ± 1.67
FSI (12 months)	51.06 ± 16.81	53.42 ± 18.27	52.24 ± 16.55	54.42 ± 1261

Note: Percentages may not total to 100% due to missing information or values.

majority. Most subjects were covered by some form of insurance; few subjects were indigent. Lastly, all four samples contained greater numbers of high school graduates and persons who had either attended or graduated from post-secondary schools compared to the number of persons who had not attended high school.

Mean SIP Scores

Mean scores ranged between 12.75 and 22.08 for the total SIP, 12.36 and 17.78 for the psychosocial dimension, and 7.64 and 19.30 for the physical dimension. In each case, the SCI sample held the highest average values and was the only sample where the mean score indicates the presence of severe dysfunction. One-way analysis of variance procedures were used to determine whether significant differences existed between the mean SIP scores of the samples. The results of these procedures indicate that with a significance level of 0.05, significant differences, $F(3, 459) = 11.69$, $p = 0.000$, were found only for contrasts containing the mean score of the SCI sample. The mean physical scores of this sample were significantly different (Tukey a, $p < 0.001$) from those of all of the other samples (SCIs, x = 19.30 ± 12.5 ; TBIs, x = 8.93 ± 11.90, IAFx, x = 10.98 ± 12.16; and burns, x = 7.64 ± 12.30) and the sample's mean overall SIP score was significantly different from those of IAFx and burn samples (Tukey a, $p < 0.001$, respectively), but not from the score of TBI sample (SCIs, x=22.08 ± 15.32; TBIs, x = 15.66 ± 5.99, IAFx, x = 13.918 ± 15.88; and burns, x = 12.75 ± 17.90). All differences between the mean score pairings for the psychosocial dimension were non-significant, $F(3, 459) = 2.42$, $p = 0.066$ (SCIs, x = 17.78 ± 18.64 ; TBIs, x = 17.61 ± 9.94 IAFx, x = 12.36 ± 17.16; and burns, x = 13.51 ± 19.63).

Simple Bivariate Correlations

To establish the existence, strength, and direction of associations between the independent and dependent variables, Pearson's correlation coefficients are displayed in Table 2. This table contains the zero-order correlations between the independent variables and the score for the overall SIP at two years after discharge. Significant correlates were as follows:

(1) SCI – discharge disposition, marital status at 12 months, employment status, level of family satisfaction, injury severity, race, and. education.
(2) TBI – discharge disposition, employment status at 12 months, marital status at 12 months, insurance status, education, level of family satisfaction, injury severity, whether or not rehabilitation was received, and whether or not a rehabilitation specialist was consulted.
(3) IAFX – race, gender, marital status, employment status, insurance status, and education.
(4) Burns – intentionality, marital status, employment status, injury severity, and insurance status.

Table 2. Zero Order Correlations Between Variables in the Model and the Total SIP.

Variable	Injury type			
	SCI	TBI	IAFx	Burns
Age at onset	−0.154	0.026	−0.114	0.039
Race	−0.294	−0.158	−0.275	−0.136
Gender	0.090	0.041	−0.176*	0.170
Alcohol	0.135	0.034	0.008	0.059
Discharge disposition.	0.327*	0.148*	0.029	−0.109
Intentionality	0.170	0.017	0.039	0.431***
Abbreviated injury scale score	−0.400**	0.002	0.079	0.058
Marital status	−0.254*	−0.131*	−0.257**	−0.336**
Employment status	−0.344* *	−0.372***	−0.379***	−0.440***
Blame	0.150	−0.103	−0.098	0.070
Education	−0.314*	−0.203**	−0.291**	−0.091
Family satisfaction	−0.315*	-0.305***	−0.133	−0.110
Specific severity	0.309*	0.088	0.133	0.274**
Rehabilitation received	0.090	0.235***	0.061	0.092
No insurance	0.030	0.145*	0.180*	0.216*
Rehab. physician consulted	0.209	0.132*	–	0.057

* $p < 0.05$ (one –tailed), ** $p < 0.01$ (one-tailed), *** $p < 0.001$
$^+ p < 0.05$ (two-tailed); $^{++} p < 0.01$ (two-tailed).

The largest correlations for all four samples were found between overall SIP score and employment status. In this respect, it appears that of the variables utilized in this study, employment status may be the most important predictor of scores for the overall SIP. It was most highly correlated with the overall SIP score of the burn sample ($r = -0.440$), less highly with the TBI ($r = -0.372$) and IAFx ($r = -0.379$), scores, and least highly associated with scores for the SCI sample ($r = -0.344$).

Explanatory Models and Variable Domains

The following sections contain results for the regression models explaining the health status of the four respective samples. These sections present interpretations for analyses simultaneously examining: (a) the relationships between SIP scores and the individual independent variables; and (b) the contributions of the variable domains. Results for the SCI sample are presented first, and followed sequentially by results for the remaining samples.

Table 3. Regression Models Predicting Overall Health Status: SCI Sample Variables.

Variables	β	b	t	β	b	t	β	b	t	β	b	t	β	b	t
Clinical															
Intentionality	0.114	5.747	0.809	0.115	5.838	−0.810	0.160	8.104	1.169	0.164	8.276	1.084	0.138	7.001	0.896
Injury severity	0.287	4.109	2.044*	0.242	3.457	1.584	0.203	2.902	1.320	0.082	1.176	0.492	0.099	1.414	0.595
Access to care															
Insurance status				0.047	1.501	0.329	0.101	3.240	0.734	−0.004	−0.122	−0.023	−0.018	−0.574	−0.109
Rehab. physician				0.1273	0.870	0.834	0.057	1.732	0.374	0.075	2.283	0.421	0.045	1.371	0.250
Level of care															
Discharge disp.							0.520	18.462	2.533*	0.293	10.403	1.226	0.279	9.921	1.151
Rehab. received							−0.394	−17.214	−2.012	−0.178	−7.771	−0.779	−0.162	−7.101	−0.715
Sociodemographic															
Alcohol										−0.006	−0.217	−0.039	0.027	0.936	0.159
Age at onset										−0.085	−0.084	−0.492	−0.025	−0.024	−0.138
Education										−0.253	−7.669	−1.651	−0.256	−7.760	−1.637
Employment status										−0.182	−7.952	−1.082	−0.210	−9.180	−1.251
Gender										0.068	2.346	0.437	0.085	2.932	0.532
Marital status										−0.100	−3.021	−0.562	−0.054	−1.630	−0.296
Race										−0.053	−1.826	−0.313	−0.083	−2.865	−0.490
Psychosocial															
Family satisfaction													−0.045	−0.041	−0.254
Self blame													0.234	3.158	1.580
R^2	0.108			0.122			0.240*			0.350			0.395		

* $p < 0.05$, ** $p < 0.01$, *** $p < 0.001$; one-tailed t tests
R^2 significance change (* $p < 0.05$, ** $p < 0.01$).

Table 4. Regression Models Predicting Overall Health Status: TBI Sample.

Variables	β	b	t	β	b	t	β	b	t	β	b	t	β	b	t
Clinical															
Intentionality	−0.007	−0.996	−0.100	−0.015	−0.209	−0.210	−0.013	−0.191	−0.196	−0.002	−0.031	−0.034	−0.004	−0.056	−0.063
Injury severity	0.085	1.346	1.204	0.035	0.559	0.479	−0.032	−0.511	−0.403	−0.091	−1.449	−1.178	−0.093	−1.472	−1.237
Access to care															
Insurance status				0.135	5.190	1.954	0.146	5.603	2.146*	0.072	2.758	0.997	0.091	3.492	1.300
Rehab. physician seen				0.123	7.019	1.695	0.059	3.371	0.760	0.048	2.762	0.650	0.014	0.785	0.189
Level of care															
Discharge disp.							0.028	1.021	0.311	0.016	0.574	183	0.008	0.304	0.099
Rehab. Received							0.224	7.198	2.863***	0.234	7.524	3.038**	0.237	7.627	3.183**
Sociodemographic															
Alcohol										0.001	0.027	0.010	−0.011	−0.399	−0.156
Age at onset										0.045	0.041	0.587	0.064	0.059	0.864
Education										−0.150	−5.286	−2.195*	−0.125	−4.410	−1.882
Employment status										−0.269	−8.750	−3.574**	−0.249	−8.073	−3.379**
Gender										0.0612	0.095	0.887	0.0612	0.082	0.910
Marital status										−0.044	−1.437	−0.603	−0.030	−0.989	−0.428
Race										−0.047	−1.852	−0.703	−0.036	−1.423	−0.557
Psychosocial															
Family satisfaction													−0.248	−0.218	−3.844***
Self blame													0.011	0.0782	0.169
R^2*		0.007			0.038*			0.084**			0.212***			0.270***	

*$p < 0.05$, ** $p < 0.01$, *** $p < 0.001$; one-tailed t tests
R^2 significance change ($p < 0.05$, ** $p < 0.01$)

Spinal Cord Injuries

The results of the analysis conducted in the SCI sample are presented in Table 3. The data in this table suggest that none of the explanatory variables are significantly related to the outcomes of this particular sample. The only variables attaining statistical significance at any point in the hierarchical regression procedures were injury severity and discharge disposition. These factors achieved significance at their point of entry but lost this status when the access to care and socio-demographic domains, respectively, were inserted into the regression model. The results of this group are unreliable, due to the small number of subjects relative to the large number of predictor variables used in the regression analysis. The beta-values obtained may be non-significant however they are of comparable magnitude to those in other groups.

In spite of the poor performance of individual variables, the collective model explained 40% of the variance in the SIP scores of the SCI sample. Of this percentage, 11% was contributed by the clinical domain, 1% by the access to care domain, 12% by the level of care domain, and 16% by the socio-demographic and psychosocial domains. These percentages are based on each domain's contribution to the R square after domains with higher priority have contributed their share to the explanation of overall SIP score. All ensuing statements about domain contributions are addressed similarly.

Traumatic Brain Injuries

In contrast to the results presented in the prior section, significant individual relationships were observed between SIP scores and specific independent variables (see Table 4). Three variables were clearly related to the outcomes of persons with TBIs: having received rehabilitation, employment status, and family satisfaction. Lower SIP scores, that is, better outcomes, were reported by those who had not received rehabilitation, who were employed at 12 months after injury, and who reported higher levels of family satisfaction. The first two variables achieved and maintained statistical significance despite the insertion of other factors into the regression model. This finding may highlight the salience of these variables for perceptions of overall health. The study variables collectively accounted for 27% of the variation in overall SIP scores for the TBI sample. Seven-tenths, of 1% of the variance in the SIP scores was explained by the clinical domain, 3% by the access to care domain, 5% by the level of care domain, and 19% by the sociodemographic and psychosocial domains.

Intra-articular Fractures of the Lower Extremities

The regression model for the IAFx sample (Table 5) contained the most variables involved in significant individual relationships with SIP overall scores.

Table 5. Regression Models Predicting Overall Health Status: IAFx Sample.

Variables	β	b	t	β	b	t	β	b	t	β	b	t	β	b	t
Clinical Intentionality	0.059	0.972	0.596	0.078	1.301	0.805	0.054		0.627	-0.009	-0.155	-0.097	-0.008	-0.126	-0.077
Injury severity	0.137	1.607	1.389	0.133	1.564	1.373	0.124	1.454	1.257	0.129	1.517	1.475	0.126	1.483	1.373
Access to care															
Insurance status				0.197	8.381	2.037*	0.210	8.948	2.125	0.095	4.053	0.970	0.094	3.988	0.931
Level of care															
Discharge disp.							-0.017	-1.010	-0.168	-0.016	-0.089	-0.016	0.000	0.0005	0.000
Rehab. received							0.077	2.548	0.763		3.558	1.181	0.110	3.656	1.188
Sociodemographic															
Alcohol										0.095	-4.994	0.964	0.095	-4.961	0.942
Age at onset										0.240	0.219	-2.413*	0.240	0.219	-2.390*
Education										0.219	-8.031	-2.383*	0.222	-8.134	-2.367*
Employment status										0.326	-11.001	-3.368**	0.320	-10.807	-3.159**
Gender										0.044	-1.417	0.473	0.046	-1.458	0.481
Marital status										0.079	-2.530	0.798	0.078	-2.50	2.759
Race										0.176	-6.635	-1.819	0.179	-6.741	-1.809
Psychosocial															
Family satisfaction													0.021	0.022	0.222
Self blame													0.007	0.074	0.070
R^{2}*	0.020.			0.058*			0.064.			0.334***			0.334		

* $p < 0.05$, ** $p < 0.01$, *** $p < 0.001$; one-tailed t tests

* R^{2} significance change (* $p < 0.05$, ** $p < 0.01$).

It shares a common predictor with both the TBI and burn samples (see Tables 4, 5, and 6): employment status. However, it differs from these samples in that relationships involving the variables age at onset and education were also significant. Whereas lower SIP scores were found among persons who were older, more educated, and employed, higher overall SIP scores were found among individuals who were younger, less educated, and jobless. The collective model for this injury sample explained 33% of the variance in overall SIP score, the majority of which was contributed by the sociodemographic domain. This domain explained 27% of the variance in the SIP scores, whereas the clinical, access to care, and level of care domains respectively explained 2%, 4%, and 1% of the score variation. Finally, the psychosocial domain did not explain any of the variance in the outcome of the IAFx sample.

Burns

Two independent variables shared significant relationships with the overall SIP score of the final sample (see Table 6). Intentionality and employment status were individually related to the outcome of the burn sample. Higher SIP scores were observed among the intentionally injured and the unemployed, and lower SIP scores were observed for the employed and for persons with unintentional injuries. Collectively, the study variables explained 44% of the variation in the burn sample's scores. Clinical variables explained 24% of the variance, access to care and level of care variables each explained an additional 2%, and the socio-demographic and psychosocial variables explained 15%.

DISCUSSION

The current study was devised to: (a) measure and compare the health status of persons with SCIs, TBIs, burns, and IAFx; (b) identify factors influencing variation in the health of persons with each injury type; and (c) determine whether the effects of study factors on health are consistent across injury types. The first of these objectives was fulfilled by using the SIP to indicate the central outcome measure. The remaining objectives were achieved by developing and statistically evaluating a series of literature-based hypotheses. Although the data did not support all of our predictions, factors associated with health status were determined, and the relative contributions of the dimensions were characterized. The following sections contain discussions corresponding to findings for each of the aforementioned objectives.

Table 6. Regression Models Predicting Overall Health Status: Burns Sample.

Variables	β	b	t	β	b	t	β	b	t	β	b	t	β	b	t
Clinical Intentionality	0.418	36.131	4.471***	0.418	36.133	4.484***	0.418	36.105	4.450***	0.362	31.274	30.946***	0.343	29.659	30.677***
Injury severity	0.234	4.370	2.500*	0.198	3.701	2.061*	0.180	3.360	1.819	0.118	2.203	1.136	0.135	2.518	1.282
Access to care															
Insurance status				0.140	6.689	1.453	0.160	7.653	1.628	0.061	2.913	0.584	0.091	4.361	0.851
Rehab. physician				0.089	7.730	0.957	0.091	7.858	0.965	0.058	5.026	0.628	0.050	4.336	0.539
Level of care															
Discharge disp.							−0.078	−4.110	−0.836	−0.126	−6.625	−1.313	−0.132	−6.937	−1.350
Rehab. Received							0.108	3.977	1.121	0.144	5.280	1.534	0.110	4.046	1.130
Sociodemographic															
Alcohol										0.061	30.789	0.627	0.041	20.576	0.419
Age at onset										0.025	0.028	0.238	0.028	0.031	0.258
Education										−0.031	−1.143	−0.323	−0.007	−0.255	−0.071
Employment status										−0.260	−9.372	−2.265*	−0.267	−9.620	−2.308*
Gender										−0.059	−2.680	−0.565	−0.067	−3.048	−0.636
Marital status										−0.198	−7.277	−1.805	−0.177	−6.495	−1.589
Race										0.020	0.840	0.203	−0.006	−0.265	−0.062
Psychosocial															
Family satisfaction													−0.054	−0.077	−0.580
Self blame													0.115	1.467	1.198
R^2	0.244***			0.267			0.284			0.421*			0.434*		

* $p < 0.05$, ** $p < 0.01$, *** $p < 0.001$; one-tailed t tests

* R^2 significance change (* $p < 0.05$, ** $p < 0.01$)

Levels of Health Status

Health status assessments obtained at two years after discharge from acute care showed that injury-related problems persisted despite restorative and rehabilitative efforts. The SCI group presented the most unique functional profile of the four injuries. On average, persons with this injury type showed markedly worse overall functioning than persons having IAFx and burns. The SCI sample's distinctions in these areas stem from the presence of the highest levels of dysfunction across the physical and psychosocial domains of health. The levels of dysfunction observed in the performance of the activities included in these domains are consistent with those cited elsewhere (see Cole et al., 1991; Stover & Fine, 1986) and could be expected given the extensive impact of spinal cord injuries upon motor and sensory function. Although not statistically significant, the health states of individuals with TBIs were generally better than the states of individuals with SCIs and poorer than those of individuals with IAFx and burns. Compared with the SCI and IAFx samples, physical dysfunction for the TBI sample was found to be minimal. Its dysfunctions in this area mirrored those of the burn sample. Persons with burns were the least physically impaired at the time of study, evidencing functional levels higher than those reported by all other subjects. This fact seems logical given that this sample also held the lowest severity scores and, by extension, possessed the mildest injuries. The moderate dysfunction levels among those with IAFx, were, however, unexpected because the literature suggests that good functional recovery in this population has often been achieved either at or before this point (Butcher et al., 1996; Mackenzie et al., 1987).

In considering the discussion presented above, it is important to note that no attempts were made to control for the effects of conditions secondary to the initial injuries. Previous research indicates that subjective assessments of health status may be jointly affected by the coexistence of primary and secondary conditions. In reference to the current study, this suggestion makes it evident that although traumatic injuries are independently associated with poor functional evaluations, even poorer evaluations may be obtained when these factors are coupled with complications such as pressure sores, spasticity, and severe depression. The apparent effects of the injuries on health status may thus be greater among persons having secondary conditions than among persons without these conditions. The presence of such conditions may deflate health perceptions and subsequently lead to the acquisition of either similar or dissimilar reports of health status from the different injury samples. Secondary conditions may, for instance, cause persons with TBIs and IAFx to report similar levels of psychosocial health, and persons with SCIs and burns to report significantly different levels of physical health. The validity of this line of thinking must be evaluated in future research comparatively examining the relationship between injury and health.

Significant Determinants of Overall Health Status

In keeping with our initial expectations, the data demonstrated that the factors related to better overall health status included: being employed for persons with TBIs, IAFx and burns; having higher levels of family satisfaction for persons with TBIs; and possessing a high school level education or above for persons with fractures. Two factors were linked to diminished health status: having received rehabilitation for TBI patients, and possessing intentional injuries for burn patients. Regarding the determinants of health status at two years after hospital discharge, explanatory models for each injury type differed in terms of which variables attained significance and their contributions to explaining variations in the outcomes. The one factor commonly shared by three of the injury samples was a sociodemographic characteristic: employment status. This variable attained significance in three of the four hierarchical regression models.

Past research has shown that employment strongly promotes positive outcomes by contributing to one's sense of well-being and degree of social integration, facilitating self-efficacy, and increasing the likelihood that an individual will have insurance coverage (Horn, Yoels, Wallace, Macrina & Wrigley, 1998; Lundqvist et al., 1991; O'Neill et al., 1988; Warren et al., 1996; Webb et al., 1995). It is likely that health status reflects the operation of similar underlying dynamics. It is also possible that employment positively influences health by reducing the level of role dissonance remaining after the onset of injury. Individuals who are able to maintain or acquire employment despite their injuries are more likely to define their health more positively than individuals who have not achieved or maintained such accomplishments. In this respect, differences in perceptions of health status may partially result from the knowledge that one is or is not in compliance with society's expectations regarding employment.

There were two additional important sociodemographic determinants: education and age at onset. Education has been described as the strongest single predictor of good health (Cockerham, 1995). Greater levels of educational attainment are generally associated with higher incomes and the possession of a more extensive knowledge of the benefits of follow-up treatments such as rehabilitation. This combination of factors makes it more likely that these individuals will seek and attain the medical, psychiatric, and social services needed to resolve the problems accompanying traumatic injury. Furthermore, there may be differences in the types of social support given to injured persons with higher and lower levels of education. Well-educated persons with IAFx may receive higher levels of informational support than IAFx patients with lower educational levels. This form of support is characterized by the provision of professional information or advice that might better prepare the former group to confront both physical and psychosocial obstacles. Finally, the data showed that education was not an important determinant of the overall health status of persons with SCIs, TBIs, or burns.

One of the study's most surprising findings concerned the third and final significant sociodemographic variable: age at onset. This variable achieved statistical significance in the final model for the IAFx sample; however, the association did not assume the hypothesized positive direction. The regression analyses indicated that older persons displayed significantly better outcomes than younger persons. A potential justification for this finding is that subjective measures of health, such as the SIP, may be influenced by the tendency of elderly persons to rate their health status as good despite the fact that substantial deteriorations may have taken place. Studies addressing this tendency suggest that such ratings are influenced by the propensity of older persons to judge themselves relative to their peers and in relation to the low functional levels that society generally expects them to maintain (Myles, 1978; Stoller, 1984). From this perspective, the differences in overall health assessments of the older and younger subjects with IAFx may simply reflect different points of reference during the process of self-evaluation.

Analyses for the psychosocial variables revealed that family satisfaction is an important determinant of better health status only among persons with TBIs. High levels of family satisfaction frequently indicate that the families of persons with TBI have resolved the crisis caused by the injury in a positive manner. Such positive resolutions are often the result of increased family cooperation and the development of stronger family relationships. The support of the family may, as Gore (1985) suggests, enhance the health of TBI survivors by positively altering their perceptions of and responses to their situations. Accordingly, TBI survivors who are highly satisfied with their family relationships may view their health status more optimistically than less satisfied survivors. They may also assume more active roles in attempts at normalization. Either one of these constructive actions might be responsible for the differences in health status observed in the present study.

Only two variables were significantly associated with poorer health status: intentionality for persons with burns and having received rehabilitation for persons with TBIs. With respect to the first relationship, earlier research hypothesized that a negative labeling phenomenon (Blumer, 1969) may be responsible for reductions in the probability that persons with intentional injuries will be referred to rehabilitation (Horn et al., 1998; Wrigley et al., 1994). Such labeling may limit the intentionally injured from being referred to those with the rehabilitative skills to help them re-achieve normality, which increases the likelihood that they will be forced into permanent physiological and social deviance. Conversely, it is assumed that persons with unintentional injuries are not exposed to this form of negative social labeling. These individuals are, as a result, presented with more opportunities to receive rehabilitation and better prospects for achieving more extensive functional recoveries. Within this study, this labeling hypothesis appears to be applicable only to persons with burns.

As for the second and final significant relationship, the results indicated that that the overall health states of persons with TBIs who had received rehabilitation were significantly worse than those of individuals with TBIs who had not received rehabilitation. Previous UAB-ICRC research suggests that the effects of time must be factored into attempts to interpret findings such as these. Addressing similar findings, Webb et al. (1994, p. 1118) indicated that "given that recovery from TBI is increasingly viewed as requiring a protracted period of time, it is likely the case that 24 months post-injury is an inadequate period for measuring the long term effects of significant rehabilitation efforts." This explanation can also be used to account for the direction of the relationship between rehabilitation and health status observed in this study. It may also be the case that the more severely injured are more likely to be referred.

Comparisons of Domain Importance

Attention must be given to the relationship between the expected and observed explanatory contributions of the clinical, access to care, level of care, sociodemographic, and psychosocial domains now that the factors related to the health status of the injury samples have been identified. The clinical domain was expected to provide the most substantial contribution to the explanation of overall health because the variables comprised by this domain are indicative of the nature of the etiology of the particular injuries and the extent to which physiological functioning was compromised by the onset of injury (factors that are particularly important during the initial stages of the recovery process). However, this expectation was satisfied only in results for the burn sample, where the clinical domain explained a sizeable 24% of the outcome's variance. The limited explanatory value of the clinical domain across injury samples may perhaps be due to the limitations of the variables constituting this domain.. Only two variables were used as clinical indicators: intentionality and injury severity. Combining these two variables with other injury-related variables not addressed in the current study (e.g. variables capturing the effects of comorbidities and peri-operative complications) might lead to the creation of a more accurate representation of patients' clinical status.

Findings for the remaining samples call attention to differences in the sets of factors assuming primacy at two years post-acute care discharge. In the TBI and IAFx samples, the sociodemographic domain proved most useful in the explanation of health status, whereas the health status of the SCI sample was most clearly explained by the level of care domain. The disparities between the expected and observed contributions of the clinical domain observed in these samples show, first, that other domains may play more significant roles in determining health status and, second, that the specific domains established as most important may differ according to the type of injury sustained. Furthermore, these findings serve as reminders of the consequences of social attributes for health outcomes. In addition, the level of care received after acute-care discharge is important in establishing the extent of functioning recovered during the two-year follow-up period.

CONCLUSION

The trend towards increased survival among trauma victims that has accompanied increases in the effectiveness of life-saving technologies intensifies the need for studies of long-term outcomes. Such research must be performed if the ability of health systems to provide for and optimize the care of injury survivors is to be comprehensively evaluated. Additionally, in a period defined by increased pressures to reduce health care resource consumption, studies of long-term outcomes may help to identify injured groups at risk for experiencing extensive periods of morbidity following acute-care discharge. These groups (and their families) may require additional attention from a wider range of health care providers and greater access (or longer exposures) to rehabilitative services.

Unfortunately, America's health care system is an arena in which persons with different types of injuries must compete for funding priorities. The injury categories to be given priority are selected based on comparative assessments of the magnitude of the burdens or costs created by the defining conditions. Though a wide variety of indicators are currently used to operationalize the concepts of burden and cost, most of these indicators are purely economic in nature. For instance, three of the most commonly examined measures of morbidity costs are estimates of lost earnings, reemployment rates, and time until reemployment. These measures are important because they provide some indication of the magnitude of the public and private problems presented by injury (e.g. the effects of specific injuries on market earnings, corporate productivity, and survivors' ability to support themselves and others financially).

They are, however, representative of the "human capital" approach to valuing the indirect costs of injury, which has been criticized because it ignores dimensions of illness and injury such as quality of life (Rice & MacKenzie et. al, 1989) and does not account for the difficulties that survivors must confront during the performance of activities of a non-economic nature.

If injury groups are assigned priorities using mainly economic measures/ outcomes, the pressing needs of those whose injuries do not significantly disrupt their participation in economic activities will not be addressed. In addition, the full extent of their struggles will not be sufficiently recognized or responded to at the systemic level. Therefore, reports comparing injury groups in terms of their status on more global, humanistic outcomes are needed to supplement data on the economic impacts of different types of injuries. The present study is a first step in addressing these needs.

The global outcome that was examined in this study is health status. Previous research on health status has either been limited to comparisons of subcategories of persons with one specific form of injury, or to comparisons of no more than two different injury groups. We addressed this limitation by comparing the subjective health status assessments of persons with SCIs, TBIs, IAFx, and burns. These injuries are, arguably, four of the most significant physical life events that any person may experience.

At two years after post-acute care discharge, the overall, physical, and psychosocial health status assessments of persons with TBIs, IAFx, and burns did not differ significantly. This finding counters what many clinicians and administrators might predict based on knowledge of the differences in the sequelae associated with these injuries, and what the average person might conclude using common sense as a guiding framework. In contrast, persons with SCIs constituted a marginal group. The assessments of these individuals distinguished them from persons with the other injury types at almost every level of evaluation. These findings highlight the complexity of the relationship between injury type and subjective health status tentatively demonstrating that overall, physical, and psychosocial burdens associated with burns and IAFx are not necessarily significantly lighter than those accompanying TBIs. They also call further attention to the relative enormity of the challenges accompanying SCI and the dire need to augment services and efforts to increase the perceived and actual abilities of SCI patients to function in normal contexts.

As more effective methods for increasing the life expectancies of persons with SCIs, TBIs, IAFx, and burns are developed, the necessity of identifying health status determinants becomes increasingly apparent. Information about such variables is important because it is used to construct profiles of patients likely to need considerable health care resources, to inform patient management decisions, and to prepare the relatives of injured persons for the difficulties that might arise in subsequent years. Furthermore, this information reveals the influences that patients, families, and providers must contend with during the recovery process.

To obtain data of this type, we first identified five domains containing variables hypothesized to determine health status at two years post-discharge, established the significant determinants of overall health status for persons with each injury type, and ascertained the variable domains explaining the largest amounts of variance in each group's health status assessments. We then determined whether persons with SCIs, TBIs, IAFx, and burns could be considered distinct groups with respect to the determinants of health status. Study results revealed that the extent to which health status is explained and influenced by clinical, access to care, level of care, sociodemographic, and psychosocial variables differs according to injury type. This study extends earlier research by providing new information about the underlying patterns of similarity and dissimilarity in the functional health status of persons with four different injury types.

ACKNOWLEDGMENTS

We would like to give special thanks to Dr. Michael Wrigley, Dr. Alfred Bartolucci, Wendy Horn, and the staff of the UAB Injury Control Research Center, Birmingham, Alabama. This work was supported in part by the UAB Injury Control Research Center, Phillip R. Fine, Ph.D, MSPH, Director, and by the Centers for Disease Control and Prevention, National Center for Injury Prevention and Control, Grant R49CCR 403641.

REFERENCES

Augustinsson, L. E., Sullivan, L., & Sullivan, M. (1986). Physical, psychologic, and social function in chronic pain patients after epidural spinal electrical stimulation. *Spine, 11*, 111–119.

Bach, J. R., & Tilton, M. C. (1994). Life satisfaction and well-being measures in ventilator assisted individuals with traumatic tetraplegia. *Archives of Physical Medicine and Rehabilitation, 75*, 626–632.

Bamford, E., Grundy, D., & Russell, J. (1984). ABC of spinal cord injury: Social needs of the patient and his family. *British Medical Journal, 292*, 546–548.

Bauman, B. (1961). Diversities in conceptions of health and physical fitness. *Journal of Health and Human Behavior, 2*, 39–46.

Bergner, M., Bobbitt, R. A., Kressel, S., Pollard, W. E., Gilson, B. S., & Morris, J. R. (1976). The sickness impact profile: Conceptual formulation and methodology for the development of a health status measure. *International Journal of Health Services, 6*, 393–415.

Bergner, M., Pollard, W. E., Martin, D. P., & Gilson, B. S. (1976). The sickness impact profile: Validation of a health status measure. *Medical Care, 14*, 57–67.

Bergner, M., Bobbitt, R. A., Carter, W. B., & Gilsonm, B. S. (1981). The sickness impact profile: Development and final revision of a health status measure. *Medical Care, 19*, 787–805.

Bergner, M., & Rothman, M. L. (1987). Health status measures: An overview and guide for selection. *Annual Review of Public Health, 8*, 191–210.

Blumer, H. (1969). *Symbolic interactionism.* New Jersey: Prentice Hall, Englewood Cliffs.

Burns, R. B. (1979). *The Self Concept.* London: Longman.

Butcher, J., MacKenzie, E. J., Cushing, B., Jurkovich, G., Morris, J., Burgess, A., McAndrew, M., & Swiontkowski, M. (1996). Long-term outcomes after lower extremity trauma. *The Journal of Trauma, Injury, Infection And Critical Care, 41*, 4–9.

Caplan, A. L., Callahan, D., & Haas, J. (1987). Ethical and policy issues in rehabilitation medicine. A Hasting Center report special supplement (pp. 1–19).

Cockerham, W. C. (1995). *Medical sociology.* (6th ed.). Englewood Cliffs, New Jersey: Prentice Hall.

Cole, T. M., Maynard, F. M., Graitcer, P. L., Apple, D. F., Acree, K. H., Boen, J. R., Bontke, C. F., DiTunno Jr, J. F., Fenderson, D. A., Fine, L. J., Fine, P. R., Fuhrer, M. J., Glass, D. D., Graves, W., Helm, P. A., Jaffe, K. M., Jette, A., MacKenzie, E. J., Moskowitz, J., Perry, J., Trieschmann, R. B., Waldrep, K., Williams, T. F., & Baer, K. (1991). Rehabilitation of persons with injuries. In: *Setting the National Agenda for Injury Control in the 1990s* (pp. 495–529). Position papers from the Third National Injury Control Conference (April 22 –25). U.S. Department of Health and Human Services, Washington, D.C.

Crewe, N. M. & Krause, J. S. (1988). Marital relationships and spinal cord injury. *Archives of Physical Medicine and Rehabilitation, 69*, 435–438.

De Bruin, A. F., De Witte, L. P., Stevens, F., & Diederiks, P. M. (1992). Sickness impact profile: The state of the art of a generic functional status measure. *Social Science and Medicine, 35*, 1003–1014.

Di Scala, C., Grant, C. C., Brooke, M. M. et al. (1992). Functional outcome in children with traumatic brain injury: Agreement between clinical judgement and the functional independence measure. *American Journal Physical Medicine and Rehabilitation, 71*, 145–148.

Dombovy, M. L., & Olek, A. C. (1996). Recovery and rehabilitation following traumatic brain injury. *Brain Injury, 11*, 305–318.

Elinson, J., & Siegmann, A. E. (1979). *Sociomedical Health Indicators.* Farmingdale, N. Y.: Baywood.

Elliot, T. R., Herrick, S. M., Witty, T. E., Godshall, F., & Spruell, M. (1992). Social relationships and psychosocial impairment of persons with spinal cord injury. *Psychology and Health, 7*, 55–67.

Fisher, D. C., Lake, K. D., Reutzel, T. J., & Emery, R. W. (1995). Changes in health-related quality of life and depression in heart transplant recipients. *Journal of Heart and Lung Transplantation, 14*, 373–381.

Fuhrer, M., Rintala, D., Hart, K., Clearman, R., & Young, M. (1992). Relationship to life satisfaction to impairment, disability, and handicap among persons with spinal cord injury living in the community. *Archives of Physical Medicine and Rehabilitation, 73*, 552–557.

Re, S. (1985). Social support and styles of coping with stress. In: S. Cohen & S. L. Syme (Eds), *Social Support and Health* (pp. 263–278). New York: Academic Press.

Haas, J. F. (1988). Admission to rehabilitation centers: selection of patients. *Archives of Physical Medicine and Rehabilitation, 69*, 329–332.

Horn, W., Yoels, W., Wallace, D., Macrina, D., & M. Wrigley. (1998). Determinants of self-efficacy among persons with spinal cord injuries. *Disability and Rehabilitation, 20*, 138–141.

Hunskarr, S., & Vinsnes, A. (1991). The quality of life in women with urinary incontinence as measured by the sickness impact profile. *Journal of the American Geriatric Society, 39,* 378–382.

Jacobs, B. B., & Jacobs, L. M. (1996). Epidemiology of trauma. In: D. V. Feliciano, E. E. Moore & K. L. Mattox (Eds), *Trauma* (3rd ed., pp. 15–30). Stamford, Connecticut: Appleton and Lange.

Jurkovich, G., Mock, C., Mackenzie, E., Burgess, A., Cushing, B., de Lateur, B., McAndrew, M., Morris, J., & Swiontkowski, M. (1995). The sickness impact profile as a tool to evaluate functional outcome in trauma patients. *The Journal of Trauma: Injury, Infection, and Critical Care, 39,* 625–631.

Kaplan, H. B. (1989). Health, disease, and the social structure. In: Freeman, H. E. & S. Levine (Eds), *Handbook of Medical Sociology* (4th ed., pp. 46–68). Inglewood Cliffs, New Jersey: Prentice Hall.

Katz, S., Heiple, K. G., Downs, T. D. et al. (1967). Long-term course of 147 patients with fractures of the hip. *Surgery, Gynecology, and Obstetrics, 124,* 1219–1230.

Koska, M. T. (1988). Payment blocks access to head-injury rehab. *Hospitals, 4,* 63.

Larson, J. S. (1991). *The Measurement Of Health: Concepts And Indicators.* New York: Greenwood.

Lazar R. B., Yarkony G. M., Ortolano D. et al. (1989). Prediction of functional outcome by motor capability after spinal cord injury. *Archives of Physical Medicine and Rehabilitation, 70,* 819–822.

Lundqvist, C., Siosteen, A., Blomstrand, C., Lind, B., & Sullivant, M. (1991). Spinal cord injuries: Clinical, functional, and emotional status. *Spine, 16,* 78–83.

MacKenzie, E. J., Burgess, A. R., McAndrew, M. P., Swiontkowski, M. F., Cushing, B., deLateur B. J., Jurkovich, G. J., & Morris Jr., J. A. (1993). Patient-oriented functional outcome after unilateral lower extremity fracture. *Journal of Orthopaedic Trauma, 7,* 393–401.

MacKenzie, E., Shapiro, S., Smith, R., Siegel, J., Moody, M., & Piti, A. (1987). Factors influencing return to work following hospitalization for traumatic injury. *American Journal of Public Health, 77,* 329–334.

McDowell, I. & Newell, C. (1996). *Measuring health: A guide to rating scales and Questionnaires* (2nd ed.). New York and Oxford: Oxford University Press.

Morris, J. A., Limbrid, T. J., & MacKenzie, E. (1996). Rehabilitation of the trauma patient. In: D. V. Feliciano, E. E. Moore & K. L. Mattox (Eds), *Trauma.* (3rd ed., pp. 1013–1022). Connecticut: Appleton and Lange.

Moore, A. D., Stambrook, M., Gill, D. D., & Lubusko, A. A. (1992). Differences in long-term quality of life in married and single traumatic brain injury patients. *Canadian Journal of Rehabilitation, 6,* 89–98.

Myles, J. F. (1978). Institutionalization and sick role identification among the elderly. *American Sociological Review, 43,* 508–521.

Oliver, J. H., Ponsford, J. L., & Curran, C. A. (1996). Outcome following traumatic brain injury: A comparison between 2 and 5 years after injury. *Brain Injury, 10,* 841–848.

Olson, D. H., Sprenkle, C., & Russell C. (1979). Circumplex model of marital and family systems I: Cohesion and adaptability dimensions, family types, and clinical application. *Family Process 18,* 3–27.

O'Neil, J., Hibbard, M. R., Brown, M., Jaffe, M., Sliwinski, M., Vandergroot, D., & Weiss, M. J. (1998). The effect of employment on quality of life and community integration after traumatic brain injury. *Journal of Head Trauma Rehabilitation, 13,* 68–79.

Patterson, D. R., Questad, K. A., Boltwood M. D., Covey, M. H., de Lateur B. J., Dutcher, K. A., Heimbach, D. M., & J. A. Marvin. (1987). Patient self-reports three months after sustaining a major burn. *Journal of Burn Care and Rehabilitation, 8,* 274–279.

Ptacek, J. T., Patterson, D. R., Montgomery, B. K., & Heimbach, D. M. (1995). Pain, coping, and adjustment in patients with burns: Preliminary findings from a prospective study. *Journal of Pain and Symptom Management, 10,* 446–455.

Questad, K. A., Patterson, D. R., Boltwood, M D., Heimbach, D. M., Ducher, K. A., Lateur, B. J., Marvin, J. A., & Covey, M. H. (1988). Relating mental health and physical function at Discharge to rehabilitation status at three months postburn. *Journal of Burn Care and Rehabilitation, 9,* 87–89.

Russel, D., & Cutrona, C. (1984). The Provisions of Social Relationships and Adaptation to Stress. Paper Presented at the meeting of the American Psychological Association, Toronto, Canada (August 24–28).

Scrambler, G. (1991). *Sociology as Applied To Medicine.* London: Bailliere-Tindall.

Stambrook, M., Moore, A. D., Peters, L. C., Zubek, E., McBeath, S., & Friesen, I. C. (1991). Head injury and spinal cord injury: Differential effects on psychosocial functioning. *Journal of Clinical and Experimental Neuropsychology, 13,* 521–530.

Stoller, E. P. (1984). Self-assessments of health by the elderly: The impact of informal assistance. *Journal of Health and Social Behavior, 25,* 260–270.

Stover, S. L., & Fine, P. R. (1986). *Spinal Cord Injury: The Facts And Figures.* Birmingham, Alabama: University of Alabama at Birmingham.

Tian, Z. M., & Miranda, D. R. (1995). Quality of life after intensive care with the sickness impact profile. *Intensive Care Medicine, 21,* 422–428.

Warren, L., Wrigley, J. M., & Yoels, W. C. (1996). Factors associated withlife satisfaction among a population with neurotrauma. *Journal of Rehabilitation Research and Development, 33,* 404–408.

Webb, C. R., Wrigley, M., Yoels, W., & Fine, P. R. (1995). Explaining quality of life for persons with traumatic brain injuries two years after injury. *Archives of Physical Medicine and Rehabilitation, 76,* 1113–1119.

Wrigley, J. M., Yoels, W. C., Webb, C. R., & Fine, P. R. (1994). Social and physical factors in referral of people with traumatic brain injuries to rehabilitation. *Archives of Physical Medicine and Rehabilitation, 75,* 149–155.

Yarkony G. M., Roth, E. J., Heineman, A. W. et al. (1988). Functional skills after spinal cord injury rehabilitation: Three-year longitudinal follow-up. *Archives of Physical Medicine and Rehabilitation, 69,* 111–114.

Yeo, G., Ingram, L., Skurnick, J., & Crapo, L. (1987). Effects of a geriatric clinic on functional health and well-being of elders. *Journal of Gerontology, 42,* 252–258.

THE EXTENDED SELF: ILLNESS EXPERIENCES OF OLDER MARRIED ARTHRITIS SUFFERERS

Peri J. Ballantyne, Gillian A. Hawker and
Detelina Radoeva

ABSTRACT

A large epidemiological survey evaluating the extent of arthritis in the population aged 55+ uncovered a discrepancy between medically assessed need for and patient willingness to consider treatment involving total joint arthroplasty. In an attempt to understand this discrepancy, we conducted a qualitative study to assess patients' experiences of the disease. This paper is focused on how the quality of the marital relationship influences the everyday functioning of arthritis sufferers. Our results suggest that in addition to the individual's functional capacity, the couple's relational and functional behaviours influence the meaning of the disease and an individual's health care decision making.

INTRODUCTION

A recent epidemiological survey of the community dwelling population aged 55 and over living in two regions of Ontario, Canada uncovered a substantial

Changing Consumers and Changing Technology in Health Care and Health Care Delivery,
Volume 19, pages 261–282.

discrepancy between medically-assessed need for joint arthroplasty for hip and knee arthritis and patient willingness to consider this intervention. Study investigators concluded that among individuals with severe disease, there is under-use of arthroplasty among men and women,[1] and that the provision of extensive information about its availability and benefits did not necessarily improve patient demand or interest (Hawker et al., 2000, 2001).

From the perspective of medical specialists, this discrepancy is problematic, given that joint arthroplasty has been found to be cost-effective and efficacious in the treatment of pain and functional disability for advanced arthritis of the hip and knee (Hawker et al., 1998; Chang et al., 1996; Laing et al., 1986). The finding that patients under-utilize this treatment option may reflect both physicians' perspectives on the appropriateness of this treatment and their referral patterns, and patients' perspectives on and responses to the disease, including successful (or adequate) coping strategies. This paper is focused on the latter of these potential explanations.

An understanding of health care decision making by patients with a chronic, non-terminal condition such as arthritis requires investigations of the *illness* experience – that is, the subjective experience and meaning attached to physical conditions by patients (Susser, 1973). A focus on the patient's perspective is particularly important for understanding the management of arthritis. Its growing prevalence in an aging population suggests that it will increasingly be presented in patient encounters with physicians. For example, arthritis is one of the most common chronic conditions in the population aged 65+ (Verbrugge et al., 1991). It is estimated that 85% of Canadians will be affected by osteoarthritis by age 70 (Health Canada, 2000). Arthritis is a leading cause of permanent incapacity, and an important determinant of disability and institutionalization in elderly populations, particularly for elderly women living alone (Badley et al., 1994; Peyron & Altman, 1992).

Yet, the nature of arthritis determines that the afflicted individual usually *ages with the disease*. This long-term experience takes place within a changing social context so that the symptoms and effects of the disease may be re-interpreted as the context changes. Also, the illness may take on different meanings for the individual who has increasing experience managing its symptoms. Only by understanding a patient's unique historical account of their disease can medical practitioners understand the discrepancy we describe above.

We undertook a study to examine how individuals experience arthritis in their everyday social environments, and how they cope with symptoms – including chronic and episodic pain, and functional limitations. In another paper, we present a multi-level model of the individual, social-interactional and social-structural features of patients' lives in which the experience of arthritis and responses to it are contextualized (Ballantyne et al., 2001). The social-interactional context includes the proximal and everyday interaction of the marital couple or other co-habitants, as well as the broader social network that includes informal and formal ties that serve as social resources for the individual. For our current purposes, we limit our analysis to married respondents. Specifically, we describe the relationship between the quality of the marital relationship and an individual's perceptions and experience of arthritis, and their views of total joint arthroplasty as a treatment option.

The focus on the married couple is important. The marital relationship is important because it generally includes a long-term intimate socio-emotional relationship as well as established functional roles that are important for the identities of each member of the dyad. Marital status has long been used as a key measure of social support in health and social surveys. An abundant literature indicates that marriage, in general, is health enhancing. At any given age, married people have lower morbidity and mortality rates than unmarried people (Koskenvuo et al., 1986; Morgan, 1980; Trovato & Lauris, 1989). Explanations for the relationship between marital status and health include higher levels of material resources, lower levels of stress, and the quality and intimacy of social support among married individuals that may not be available to individuals who are not married (Wyke & Ford, 1992).

Yet, marriage can also involve costs and risks, especially in the face of adversity such as chronic, debilitating and terminal illness, particularly when the caregiver is elderly and in a compromised state of health (Rose & Bruce, 1995). Illness of one member of the couple may challenge the everyday functioning of the spousal unit, and the couple's response to the challenge will determine how the individual responds to illness. Marital status also affects identity (Askham, 1995) so that the quality of a marital relationship can mediate self-perceptions and the psychological capacity of an individual to define and respond to chronic disease.

METHODS

The methods and analysis of data are described more fully elsewhere (Ballantyne et al., 2001). Here, we analyze in-depth qualitative interviews with 16 married individuals with severe disease. The interviews were guided by a semi-structured interview schedule. The schedule allowed the interviewer to ask respondents about past and present experiences and future expectations of their disease and to gain insight into a respondent's perspective on health care decision-making for arthritis, and allowed *respondents* to move the interview in directions that were relevant to their own experiences. The interviews were in pursuit of the general question "in the context of their everyday lives, how do aging individuals' perceptions of and experiences with arthritis influence their management of the disease on a day-to-day basis?"

Sampling and Sample Characteristics

Responses from married individuals are based on a sub-sample of the original qualitative study (see Ballantyne et al., 2001). The original study included 29 individuals aged 55 and over who have arthritis, who were selected for the study based on clear inclusion criteria: following a three-stage assessment in the larger survey, they reported and were objectively assessed as having *severe arthritis*; they were medically assessed as appropriate candidates for total joint arthroplasty (tjr); and they rejected the tjr option after being given detailed information about the risks and benefits (for details see Hawker et al., 2000, 2001). Further selection was based on respondents' willingness to participate in the qualitative study, which represented a fourth contact with study investigators. The sample then, is not random, and findings may not be representative of all individuals with severe arthritis.

Seventeen of twenty-nine individuals who agreed to participate in the qualitative study were married at the time of their interview. One interview was excluded because of difficulties in transcription. The married respondents include 9 women and 7 men; primarily between the ages of 60 and 70, and representing two distinct geographic regions (a rural/small town area and a metropolitan area). These sample characteristics are outlined in Table 1.

Table 1. Summary Characteristics of Married Respondents with Severe Arthritis.

Characteristics		Total	County Sample	City Sample
Age	< 60	1	1	0
	60–70	12	11	1
	70–80	3	1	2
	> 80	0	0	0
Sex				
	Female	9	7	2
	Male	7	6	1
N		16	13	3

Analysis

All interviews were tape recorded and transcribed verbatim. Transcripts were subjected to inductive content analysis (Berg, 1995; Holstein & Gubrium, 1994; Strauss & Corbin, 1994; Glaser & Strauss, 1967). This involves multiple readings of the text of the transcripts. At first reading, central themes and major general issues were identified. Subsequent readings resulted in an open-coding scheme. This phase is complete when the codes have become "saturated", and involves repetitious reading of the text until all data is coded. Analysis of the transcripts, following this pattern of iterative review of themes, codes and sub-codes, allows the analysis to move from description to explanation, from the concrete to the more abstract, following the grounded theory methodology (Glaser & Strauss, 1967). Ethnograph software for the management of text-based data was employed for open data coding (Qualis Research, 1998).

RESULTS

Qualitative data codes that were the exclusive domain of married respondents were extracted from the codebook. Four specific codes were unique to married respondents: division of household labour, dependence on spouse, support from a spouse, and conflict with spouse. The first two describe *activities*. "Division of labour" describes a respondent's self-assessed responsibilities in terms of the maintenance of a household and household members. "Dependence on spouse", a marker of functional limitation of the respondent, refers to situations where a respondent is unable to perform a specific activity or function without the assistance of their spouse. The second two codes describe *characteristics of the interaction or relationship* between the respondent and their spouse. "Spouse support" describes the respondent's acknowledgement of *positive* interaction and assistance from the spouse for carrying out specific activities or functions. "Conflict with spouse" refers to a respondent's report of overt or latent *negative* interactions with the spouse.

It is important to note that these codes are not mutually exclusive, and their co-occurrence in the data reveals the richness of text-based analysis. The complexity of marital interaction may be expressed, for example, in a case where a respondent describes dependence on their spouse for an activity such as assisted walking outside the house, and conflict with the spouse over this arrangement. Or, one individual may describe being dependent on a spouse for the completion of some activity (implying inability to perform it independently), while another will describe receiving support from a spouse in carrying out an activity that would otherwise be very difficult to perform independently. The difference here is subtle, with these distinctions marking the degeneration of function that some experienced, as well as the strategy for coping with that degeneration, based on co-operative social interaction with the spouse. It may be the case that the functional ability of two individuals is similar, but the *perception* of their function is moderated by how the spouse serves to assist the respondent in managing a physical limitation.

In the text that follows, we describe these four coded categories more fully. Then, on the basis of our analysis of respondents' activities and interactions with the spouse, we define three distinct types of marital relationships. The relationship between these marital types and respondents' everyday experiences of arthritis, and their perceptions of joint replacement surgery as treatment, are examined.

Sharing Labour: The Division of Labour as a Social Context for Married Individuals with Arthritis

Of 16 married respondents, 14 discussed a division of labour in the household as a context in which they had to negotiate activity levels because of physical limitations related to arthritis and other co-morbidities (7 women and 7 men; 12/13 county residents and 2/3 city residents). These negotiations were with themselves, with their spouses, or with others such as friends or formal service providers. Several codes occurred simultaneous to, or as an over-lap to the division-of-labour code, providing insight into how household activities were negotiated and managed by the individual with severe-stage arthritis. These included details of the physical condition of the respondent (such as medical complications or co-morbidities, physical restrictions, functional decline, need for the use of props such as a wheelchair), the individual's psychological response to the disease state (such as resolve and a commitment to "keep going", or frustration), the social resources available (self sufficiency or reliance on self, spouse support or conflict, dependence on the spouse, social network and social exchange), responsibilities for others (caregiving), and strategies for coping with activity demands related to household labour (using restraint such as avoiding activities, using an activity control cycle, for example, will exert oneself over housework one day and refrain from exertion of any kind, the next).

Gender differences related to the management and division of household labour among individuals with arthritis were evident. Among married women respondents, 4 described a separate division of labour between the husband and wife, and 3 described a co-operative or shared division of labour. Among those describing separate roles, simultaneous coded responses included self sufficiency and resolve (referring to the individual's commitment to their role in household activities), use of restraint in carrying out activities, recognition/acceptance of functional decline and physical restrictions that limited activities related to household roles, and spouse support.[2] Among those describing co-operative or synchronous roles, simultaneous codes included spouse support and dependence-on-the-spouse, and social exchange in the context of work shared with the spouse.

Among the married men, 5 described a co-operative division of labour in the household (shared roles related to household work), 1 was physically dependent on the spouse for most activities, and 1 described a separate, gendered division of labour. This latter individual described reliance on his adult children and a broader network of friends to assist him in maintaining *his* responsibility for work outside the home (while his wife retained responsibility for work inside the house – cooking, cleaning, decorating and maintenance, etc), and he stated he had, and was able to purchase formal assistance, in order to continue *his* "usual responsibilities".

Gender differences in respondents' discussions of the division of household labour were revealed in another way as well. While only 2 of 16 respondents described care-giving responsibilities as an aspect of their household work, both respondents were women (Another male respondent discussed *his wife's* ongoing responsibility for continuous care to him). Despite their own physical health problems and functional limitations, the two women discussed their commitment to care-giving provided to adult children (one woman discussed past caregiving of an adult son with mental illness, and another woman discussed her strategies to minimize the burden of care-giving to her adult son who she viewed as having significant support needs related to his poor health, broken marriage, and parenting responsibilities). The husband of this latter woman was also highly dependent on her because of his poor health. She included the care of her husband as part of her everyday household labour, and she acknowledged that the burden of care was difficult because of her own state of health. Importantly, she demonstrated how she did successfully find some relief:

R: ... I, huh, picked up my husband a few times already too, but I'm getting smart. That, too, I ... I've learned how to make him know – I've got to learn HIM to do it. Because if I'm not here ...

I: You can't do everything for him.

R: So, I says, "Come on, do it". So the shower with – he's accomplishing more and more all the time. And he's really getting particular now, he wants this kind of clothes and, ah, gee, you know, you're really getting back to normal, you know. But he did ... he does his toes, and the only thing I do is maybe the back to make sure it's dry, you know, and ah, he dresses himself now. Used to be I had to help him all the time.

Differences in the division of household labour by regions (small town/rural and urban) are notable. In total, only 2 of 3 married city dwellers discussed the division of labour as an aspect of life they had to negotiate because of their arthritis condition, while 12 of 13 county dwellers discussed the division of labour. The city dwellers are both men, living with their wives in high-rise apartment buildings. For these two couples, "outside work", usually described as the domain of men in the county sample, was not an issue at all. Among the county residents who discussed the division of labour between themselves and their spouse, 12 lived in houses; only one lived in a retirement residence, and did not have responsibility for "outside" work. This difference will explain, in part, the stronger presence of a gendered division of labour among the county residents.

With respect to respondents' discussion of the negotiation of the household division of labour, it appears that rather than the disease curtailing (household maintenance) activities, respondents' attachment to a gendered division of labour may assist arthritis sufferers in keeping the negative aspects of the "*disease*" label at bay. The maintenance of normal, valued social and household activities may be identity-confirming – the maintenance of these valued roles no doubt assists the individual to feel continuity with their past self. While the *performance* of activities may be modified because of physical status, respondents revealed a reluctance to give up valued or usual social roles; as if this implied *giving in* to the disease.

Dependence on Spouse and Conflict with Spouse

Of 16 married respondents, 6 described being dependent on their spouse for assistance with daily activities (3 women and 3 men; 4 from the county sample and 2 from the city sample). "Dependence on the spouse" indicated that an activity couldn't be performed without the spouse's assistance, and so referred to activities related to specific tasks, as well as to an overall state of dependence and loss of self. (In this data set, two individuals had physical limitations so extensive that they were totally dependent on their spouses, and described themselves in this way.) Dependence on the spouse was related to the presence of physical restrictions and functional limitations related to (an) illness(es).

Dependence for assistance with a specific activity could be offset when reciprocity was retained by a fair or usual balance of exchange in tasks and interactions in general. In discussions related to dependence on a spouse (in general, and related to the specific activities), the simultaneous experience of conflict with the spouse was indicated by several women. These two themes are discussed below.

Overlapping codes for those who describe some level of dependence reveal differences between how these women and men feel about their dependence. For women, codes that overlapped with "dependence on spouse" included having physical restrictions, division of labour (so their dependence is related to difficulties in the performance of some self-defined household task), feeling supported by their spouse over their dependence needs, desire for formal services and scarcity of money to purchase such services, and conflict – for example, over negotiating to get assistance with specific tasks from their husbands). One woman who reported conflict over the fact of her dependence and the manner in which her husband provided assistance, at a different point in the interview also remarked on how supportive her husband had been – so being in conflict over interacting with a spouse occurred in the same context where the spouse's support was acknowledged. Two women who described being in conflict with their spouse stated it was over the fact that their husbands didn't want to talk about, or hear them "complain" about their health problems. Another's husband had had a stroke and was no longer able to share a companionable relationship (where the two could share in conversation) such as they had had in the past. Another woman – highly physically dependent for daily care from her husband, experienced conflict with him around a myriad of issues: her physical care such as mobility, toileting, eating of meals, her interaction with others (family, health care aid), and her relationship with him.

In contrast, none of the men who described situations of dependence on their spouse described conflict over the arrangements. Three males reported dependence based on physical restrictions related to having arthritis and other health conditions, their need for exercise-therapy and the completion of household activities that they defined themselves as responsible for. Simultaneous codes for these men included acknowledgement of the support provided by the spouse, their psychological response of resignation to their arthritis condition, and their resolve to continue to attempt to cope as best they could, and even their feelings of being self sufficient (*as a couple*) in meeting their needs. So in general, dependence appeared less problematic for men than for women.

In terms of regional differences in response to being physically dependent, one county resident described her reliance on her spouse because she was unable to climb the stairs in her house to fetch stores, etc. Two other (female) county residents described reliance on the spouse for driving. The two city dwellers who described being dependent were not constrained by the location or type of residence, but rather by their poor health – including significant co-morbidities, in addition to severe arthritis.

Suppport from Spouse

Of 16 married respondents, 13 acknowledged that their ability to cope was due, at least in part, to the support provided by their spouse (6 women and 7 men; 10 county residents and 3 city dwellers). "Spouse support" describes the respondent's acknowledgement of *positive* interaction and assistance from the spouse for carrying out specific activities or functions. This category is subtly different than "dependence on spouse" because the respondent describes the *ability to perform activities and function because of* support received, rather than *an inability to perform an activity* or function.

Discussions illustrating support are related to several topics, including fulfilling one's responsibilities related to the division of labour; managing physical restrictions and medical complications related to co-morbidities, maintaining a workable psychological orientation to illness (maintaining resolve, managing uncertainty, maintaining a sense of self sufficiency), a moral philosophy of responsibility to "keep going", having ongoing social interaction with friends and family, and availability of the spouse as confidante as well as the provider of instrumental support.

Support from a spouse was not always viewed as unproblematic, and as we note in the discussion of conflict, above, gender is not irrelevant here. For example, we mentioned previously, the case where a woman who is functionally dependent on her spouse for most tasks acknowledged both supportive (positive) interaction with him, but also serious conflict in some interactions. Additionally, one female county resident, who was relatively young – aged 63 at the time of the interview, indicated that while she has a co-operative relationship with her spouse and her needs were being met at the time, her future is uncertain. She states:

R: well, if . . . if, ah, well right now I'm managing; I'm doing alright, like I
     mean I have him, but . . . if I was left alone that might be a different situ-
     ation. That's not a nice way to talk, but I mean you have to be realistic
     about things. Like he's four years older than I am. But then again, he helps.
     He says (whispers) 'I'm gonna outlive you' (chuckles)."

Finally, it is notable that 10 of the 13 respondents who discuss supportive exchanges with their spouse as an aspect of living with severe arthritis are county residents. Most of these individuals continue to live in houses, maintain yards and gardens and vehicles. Supportive interaction with the spouse is one of the ways in which independent living (and for some, independent, gendered roles) is maintained.

Our discussion of these four codes – division of labour, dependence on spouse, support from spouse, and conflict with spouse – highlight the ways in which respondents experience arthritis in the context of their marriages. "Division of labour" and "dependence on spouse" describe respondents' negotiations over the completion of *activities*. These activities relate primarily to self-defined social roles in the household division of labour, and for some women, in caregiving work.[3] "Spouse support" and "conflict with spouse" describe respondents' perceptions of their everyday *interactions* with their spouses (as positive, ie: supportive; or as problematic, ie: conflict) – that is, the *relational* aspects of coping with arthritis, within the context of marriage.

On the basis of these codes, we apply three concepts to describe variations in the quality and function of respondents' marital relationships. These include synchronous relationships, independence relationships, and dependence relationships. The relationship between these marital types and respondents' everyday experiences of arthritis, and their perceptions of joint replacement surgery are summarized below.

The Synchronous Marital Relationship

Three of nine women respondents and five of eight male respondents are in marital relationships that we define as "synchronous". Synchrony, in these cases, includes both the sharing of activities as well as a shared state of mind. In sharing activities, the wife and husband may be attached to specific, and even traditional (gender-divided) roles, but one individual will act to compensate for the other's functional limitations, they will do this in a supportive manner. This compensatory-sharing will be viewed as unproblematic; as an aspect of the marriage that evolved as the couple aged. The following excerpt illustrates this view of shared activities:

I: Is there some division of labour or do you do most of the stuff, between you and your husband?
R: Ahm, well as far as the housework . . . is concerned, I do.
I: Everything?
R: . . . He does a lot of other things. Like he does, ahm, he does everything outside except my flower garden and I do that. I make – I love my flowers and I just get out and do it. I pay up for it, but I won't give it up. Because I figure it helps.

I: And it's pleasure.

R: That's right. Yeah ...

I: Yeah. Can you climb stairs easily?

R: Ah, I can – yes, I can climb them, but I can't, ahm ...

I: Go down.

R: Yes I can do it, but I can't do if like three or four times. If I do it three or four times in a day, I pay up – I really pay up for it. So I usually – like I've already been down once today; I will – if I have to go down once more, but I won't, you know ... I can't.

I: Yeah. So who does the laundry?

R: Oh, it's up on my main floor.

I: Oh it's up, okay.

R: And that – I had that, yeah, yeah, yeah.

I: This is taken care of.

R: That's taken care of. So lots of times there's no reason for me to go down there. Ah, the only thing is my freezer and sometimes I ask him to get me stuff out of it.

I: Is it easy or do you need to bargain or is he responsive?

R: Oh yeah, yeah, he's fine, yeah, yeah, yeah. And ah, I have a fruit cellar but now I've got everything out of there except a bit of canned stuff, but ahm, what I'm trying to do is when I go downstairs for one thing, I try to bring everything that I need up at the same time.

Synchrony, as we describe it here, allows (household) role-continuity, and thus serves to preserve the identity of an individual whose physical state of health is compromised. The broader text from the interview above reveals the respondent's longstanding pattern of negotiation with her spouse that continues to the present time. Another respondent's (a male) discussion of his interaction with his spouse illustrates how the couple works to support one another. It is significant here that this man's wife also has functional limitations related to her own physical health status:

I: Well, how do you handle the housework?

R: We do it between the two of us. If we don't do it today it gets done tomorrow or some other day. Don't worry too much about it.

I: Who does the groceries, or how do you do it?

R: Both of us. We go-we go with the Wheel-Trans. We take the Wheel-Trans, (coughs) that way she doesn't have to do much walking and I have this wheelchair anyway.

Another man, with a wife who also has arthritis (the wife has recently had two knee replacements), describes how he acts to compensate for his wife's functional limitations; while at the same time, recognizing his own limitations:

I: So what do you do now? Can you briefly describe for me one day, how does it go? You wake up in the morning and then what do you do? A typical day.

R: Not a whole lot. It's ... whatever housework has to be done, try and do that and cook the meals and stuff and ... the washing has to be done and do the washing, she doesn't – she hasn't done that for a couple of years, two or three years now. I can go up and down the stairs but it's not the greatest thing in the world to do. And (clears throat) you know, clean the house, vacuum, make the beds and she'd make the beds, she's starting to make the beds lately. But for a while there she wouldn't (unclear).

R: If we go for groceries, like we go for groceries ear – early in the morning is my best time. Like after resting the night, early in the morning is my best time. We'll go for groceries and go around the store and we'll maybe (wife) will sit down and have a coffee somewhere, rest there for a while – and come home and ... put the groceries away and I'll lay down for a while and you know, just anything that ...

I: Yeah, yeah.

R: Some days I can't do anything. Some days I won't do anything.

Finally, in a conversation about everyday coping, another man expressed a psychological/emotional attachment to his wife, in his everyday battle to "keep going", in both a physical and emotional way:

I: Do you walk?

R: Oh yeah, I walk from here, the garbage chute's away down the other end of the building; I take the walker and I walk around a bit.

I: Okay, yeah ...This is good ...

R: Oh yeah, I'm not gonna give up that quick. No way! ... Well especially if ... if there've been two people and one dies, the other one gives up altogether. But we got each other (speaking of wife) and that there keeps us going. I think that's what helps a lot really.

It is notable that in the above example, and in other transcripts from interviews with individuals in synchronous marriages, it was typical to see references to "me" or "I" replaced with references to "we" or "us".

Referring specifically to the respondent's assessments of total joint replacement (tjr) as a strategy for coping with their condition, among those that we describe as having synchronous marital relations, four indicated that they would be willing to undergo trj in the future (that, contrary to the medical assessments conducted earlier, they viewed trj as premature at the point of our interview, and several defined trj as potentially appropriate or necessary in the future). Four others in "synchronous" relationships were reluctant to consider tjr, even as an eventuality. Reluctance was based on reasoned consideration of past (negative) medical and hospital experience (Ballantyne, 2000) and of complicating health conditions that the individuals themselves viewed as contraindications to surgery (such as overweight). It may also be the case that *the couple* were coping sufficiently well to preclude consideration of an invasive procedure such as joint arthroplasty.

Our discussion of marital synchrony, as reflected in interactions around the completion of activities as well as at the psycho-emotional or relational level, illustrates the notion of the "extended self". The study of the *self* is concerned with the implications of the human capacity to reflect upon one's own physical, social and psychic state. Further, the self-concept is influenced by an individual's evaluation of how others view them, and how they view themselves according to socially defined rules and expectations (Mead, 1956). On the basis of our analysis, we suggest that, for some individuals, an intact sense of self can be maintained in a context of synchronous marital interactions – those that accommodate desired or usual social roles, and those with positive and reinforcing social-emotional relations between the members of the couple. The concept of "the extended self" conveys the idea that an individual may conceptualise functional limitations in the context of their most intimate relationships so that it may not be the *individual's* functional capacity, but the *married couple's* capacity to function interdependently that determines the meaning of the disease and an individual's health care decision making.

The Independent Marital Relationship

This marital type describes a relationship characterized by separate and distinct social and emotional spaces for the members of the couple. Typically, these arrangements are mutual and functional – the result of habits of interaction established between the couple long ago. In the following example, the discussion is focused on the option of joint surgery:

R: Ahm, but I just don't like the idea of not being up (chuckling), being handicapped for three weeks or four weeks or whatever it is.

I: Well if in that case, is there anybody who could help you, you know, go through this period? Because it's only for a while.

R: Well, my husband might have to. Ahm, he's not a nurturer.

I It is a type of personality ...

R: Yeah, yeah, yes, his mother did everything and he wasn't expected to do anything – well they lived – he was raised on a farm, you see, so that manual work is fine but ... Yeah, he will hel – he helps me if I ... I have a girl come in, ahm, every other week to help with house cleaning. She's helping me with the gardening too.

I: So otherwise you do the housework; or do you share things with him?

R: No-o, no (laughs). I clean up after him.

I: Okay. So if you needed some help anyway, would he need a lot of, you know, (unclear)

R: Oh no. No, no, no, he'd be, no he would help me. Yeah.

I: Okay. Just not his style.

R: No. No, he'd rather not, but ah, no – he won't do it unless I ask him, but if I ask him he'll do it.

Another respondent indicates a mutual arrangement of separation of roles, between he and his spouse (it is his friends who assist him in maintaining "his side" of the arrangement);

I: And has it been hard for you because you run a house, you and your wife live in this house and run the house; has there been any indication that you won't be able to keep the house up because of your concern or your wife's concern about your health?

R: I don't think so, because she's quite able to do anything that ... about the house if I couldn't do any, she could do it herself.

I: Has there been an exchange of roles to any significant degree ...

R: No.

I: ... in that sense where she's had to take on more?

R: No, she's always done everything in the house and I do everything outside.

I: And you pretty much keep that up now?

R: Yeah.

I: You just take the time you have to do it, and you have more time to do it now?

R: Well, when we had the snow, big snow storm, a friend of mine, he come up and he said, "Don't you be snow-blowing", so he came up and blow my snow. I got friends who can do that ...

However, aging, illness and disability may render long-established patterns of independent roles cumbersome, and even deleterious, for one member of the couple. For example, in the following case, the respondent described long-standing divided marital roles. Her husband's deteriorating health led to his inability to care for the "outside" work, and their decision, eventually, to move into a retirement apartment. She fiercely defends her turf as the homemaker (and now, caregiver to her husband), but has had to confront the collision of their previously separated worlds:

R: And he'd never seen a lot of me. And now ... Because he's never seen me. I says, "I've been like this for ten, twenty years, X, you're just not around. Like I'd be laying upstairs or something, you know, laying down and he wouldn't pay, you know, 'oh, the patient's laying down'. But now he sees that it's ...

Another woman describes a significant conflict she is currently facing, relating to housing:

I: You mentioned you were going to – you wanted to sell the house. What will you go for as a next step?
R: I'd love to go home out west, where it's nice and dry.
I: Because of the climate.
R: Um-hm.
I: Can you? ... would your family agree with that, your husband?
R: I don't know. I don't think, I don't know about X. He wants to go to a hot country and I won't go, can't stand the heat. No. I love my winters and I couldn't ... I couldn't go somewhere to be not winter (chuckles).
I: But what kind of housing situation would you choose, would that be a house again or?
R: I hate the thought of apartment living. But ... every house has some property and it has to be kept up.
I: What would you – have you thought about this? What kind of arrangement you choose next?
R: Not really, because it's just lately after years and years of trying to get him to ... say yes to selling the house, ah, he's only agreed. Now I think he's starting to back pedal because he digs his heels in about everything. But, ah, I don't know. If we can't agree we will have to go our separate ways.

This latter case illustrates that the divisions that exist between the respondent and her husband may be insurmountable, *so the marriage itself, may be a serious deterrent to her effective coping with severe arthritis.*

With respect to the joint replacement (surgical) option, among those in "independent relationships", three of five viewed tjr as a likely eventual outcome; one expressed great reluctance to consider this option, and one did not place much emphasis on the management of arthritis because of his successes at reversing functional decline (through diet, exercise, weight reduction and "a good attitude") and because of serious co-occurring morbidities. Recall that these views of the severity of the disease (and of the options for treating it) are contrary to medical assessments completed earlier in the study. And, for all respondents we categorized as in an independence-type marital relationship, none offered views on how they would eventually cope with functional loss and increased dependence. Perhaps their coping strategy could be described as avoidance.

The Dependence Relationship

Two respondents were dependent on their spouses for everyday functioning. In these cases, the capacity for reciprocity between the couple is limited. Dependence states were described in both positive and negative terms. For example, the following excerpt is from a male county resident who is highly dependent on his wife but who describes having a very positive relationship with her:

R: Yeah. Both my wife and I have retired, and we have a . . . fixed income. It's awfully hard to retire, I'll tell you . . . (laughs) Boy, it's like going from black to white, you know, or white to black, sort of thing. And it takes months upon months to . . . put in gear . . . you know, routine and everything.

I: How do you handle that?

R: Oh, well, like everything else, sorta you're not going to make anything better by . . . by struggling against it, you sorta got to go along with things; and I'm lucky that, ah, . . . that, ah, both my wife and I, we cope pretty well, luckily I got the wife that I have.

I: Oh yes. She must be helpful.

R: Oh, yes, yes she is. Yes, she looks after me like a baby, I guess (laughter). I don't go out any more when there's snow around or ice because I lost all my self-confidence since this accident, you know.

Since retirement, this man has come to live a very circumscribed existence. With his wife accompanying him on all outings, as well as caring for him in the home, he remains optimistic, and frequently expressed gratitude for his wife. This individual is unusual with respect to his optimism in view of his current (dependent) status. In most cases included in this study, the maintenance of independence, or at least the capacity for reciprocity in interactions was a central goal for everyday life (and resistance to the joint replacement option was often based on concern about even temporary dependence). In the current case, where the respondent's dependence on the spouse is described as non-problematic, the interpretation of the marital relationship as being primarily positive may reflect a reality based on a re-evaluation of self and identity-expectations, or it may reflect a process of interpretation described as marital aggrandizement (O'Rourke et al., 1998); that is, the respondent's suppression or negation of memories or beliefs that contradict a desired perception of the relationship.

Contrary to the above case, a more ambivalent view of dependence on the marital partner is evident in the following excerpt from a woman with declining health and increasing dependence on her husband:

R: I got everything wrong with me and I feel so useless . . . you know, you can't help it but . . . feel useless when you've worked all your life and done things for yourself, and then you find that you just can't do anything. You know, I mean, it's really embarrassing when you have to – he has to help me to walk, he has to help me up the stairs. If I'm going, ah, I'll walk – I'll go the wrong way. You know, it's so, oh, it's so bad. And people don't realize . . . And especially for someone who was always used to doing for myself – I've been working since I was 12; and to have to depend on somebody so totally, it's really horrible. I'd like to be able to get up and kick my heels up and you know, just do all the things other people do.

The two individuals that are functionally dependent share the view that trj is not an option at all. Their decisions are based on their own views of their health problems, of medical complications that they view as contra-indications to surgery, previous and numerous hospital experiences and an unwillingness to subject themselves to more hospital care, and uncertainty about the outcomes of tjr.

DISCUSSION

This paper is focused on the marital relationship as potentially influencing how an individual perceives and experiences the disease of arthritis. The marital relationship is important because it generally includes a long-term intimate socio-emotional relationship as well as established functional roles that are important for the identities of each member of the dyad. Illness of one member of the couple may challenge the everyday functioning of the spousal unit, and the couple's response to the challenge will determine how the individual responds to illness.

We identify three marital types: synchronous, independence and dependence relationships. Synchronous relationships involve both the sharing of activities as well as a shared state of mind. In sharing activities, the wife and husband may be attached to specific, and even traditional (gendered) roles, but one individual will act to compensate for the other's functional limitations, and they will do this in a supportive manner. This compensatory-sharing will be viewed as unproblematic; as an aspect of the marriage that evolved with time. Synchrony, as we describe it here, allows (social) role-continuity, and thus serves to preserve the identity of an individual whose physical state of health is compromised.

Our discussion of marital synchrony, as reflected in interactions around the completion of activities as well as at the psycho-emotional or relational level, illustrates the notion of the "extended self". The study of the *self* is concerned with the implications of the human capacity to reflect upon one's own physical, social and psychic state, and the self concept is influenced by how an individual views themselves according to socially defined rules and expectations. We suggest that, for some individuals, an intact sense of self can be maintained in a context of synchronous marital interactions – those that accommodate desired or usual social roles, and those with positive and reinforcing social-emotional relations between the members of the couple. The concept of "the extended self" then, conveys the idea that an individual may conceptualise functional limitations in the context of their most intimate relationships so that it may not be the *individual's* functional capacity, but the *married couple's* capacity to function interdependently that determines an individual's sense of self, the meaning of the disease and an individual's health care decision-making. This is illustrated in our interviews where individuals refer to "we" and "us" when discussing their strategies for coping with severe arthritis.

The independence relationship is characterized by separate and distinct social and emotional spaces for the members of the couple. Typically, these arrange- ments are mutual and functional – the result of habits of interaction established between the couple long ago. These arrangements, however, may become problematic and even damaging for one or both members of the couple, as they encounter illness and disability. While synchronous relationships accommodate the physical state of decline and make it manageable so that individuals delay or avoid strategies such as tjr, independence relationships are potentially problematic, at least from the perspective of continued and separate functioning. Synchrony or co-operation is not typically expressed in respondents' discussions of marital relations, and although the spouse may be accommodating, some women respondents indicated that assistance from the spouse is not easily acquired; some described having to coerce or trick spouses into assistance. In one case, insurmountable barriers to continued co-existence were indicated, suggesting that the marriage itself may be a serious deterrent to effective coping with severe arthritis.

Finally, dependence relationships, characterized by a limited capacity for reciprocity between the members of the couple due to the functional status of one, were described in both positive and negative terms. One man had ceased functioning with any independence, and acquiesced to dependence on the spouse. Recognizing the imbalance in his relationship, he nonetheless described it in very positive terms, and himself as "lucky". In the other case, the woman lamented her total loss of function and dependence on her spouse. Her frequent reference to her past self, and previous roles, were notable. To highlight the contradictions between what we described earlier as the "extended self" sustained in the synchronous marital relationship, these two cases better illustrate an "expended self" – the near total loss of function and the self as an autonomous being. It is notable that these two individuals were not dependent solely or even primarily because of arthritis, rather serious co-occurring morbidities led to their functional and emotional declines.

There are three important implications of this study. First, it should not be assumed that marriages are supportive; and it should be assumed that among aging individuals, the capacity for support within the marriage may be diminished with time. Second, individuals within marriages can identify specific needs – the maintenance of homes, gardens, the need for assistance to maintain normal social interactions, the need for assistance in preparing meals, and with personal care activities of daily living, etc. Third, where these needs cannot be met within the marital home (either because of the inability or unwillingness of the spouse to assist; or because of limited economic resources), alternative means of access ought to be pursued, in the name of health care.

NOTES

1. The study concluded there is under-use of arthroplasty for severe arthritis in both sexes, but the degree of under-use is more than three times as great in women as in men (Hawker et al., 2000).

2 An example of spouse support for the respondent's independent household role included a situation where the husband would fetch groceries for storage so that the wife could prepare meals.

3. The paper is based on discussions of the marital relationship. Text related to how respondents with severe arthritis negotiate within their broader social network is excluded

from this paper, but is also less common in the interviews. This is because of the age and life stage of these respondents (most are retired), and because for many, their advanced disease status means their everyday lives are centered primarily around activities (and non-activity) in the home, and interaction with their spouse.

ACKNOWLEDGMENTS

This study was supported by grants from the Medical Research Council of Canada (MT-12919), the Arthritis Society of Canada (97-083), Physicians' Services Foundation (95–47), the Canadian Orthopaedic Foundation, and the University of Toronto Dean's Fund (00026896) awarded to Dr. G. Hawker, Principal Investigator of the arthritis survey, and her colleagues. For completion of the qualitative study, Peri Ballantyne gratefully acknowledges support from the Centre for Research in Women's Health, Sunnybrook and Women's College Health Science Centre/University of Toronto, and the Women's College Hospital Association of Volunteers.

REFERENCES

Askham, J. (1995). The Married Lives of Older People. In: S. Arber & J. Ginn (Eds), *Connecting Gender and Ageing: A Sociological Approach* (pp. 87–97). Buckingham, U.K.: Open University Press.

Badley, E. M., Rasooly, I., & Webster, G. (1994). Relative Importance of Musculoskeletal Disorders as a Cause of Chronic Health Problems, Disability, and Health Care Utilization: Findings from the 1990 Ontario Health Survey. *Journal of Rheumatology, 21*, 15–37.

Ballantyne, P., Hawker, G., & Radoeva, D. (2001). Modelling the Factors Influencing Arthritis Experience and Decision-making Regarding Treatment. McMaster University: Social and Economic Dimensions of an Aging Population (SEDAP) Working Paper Series. http://socserv2.mcmaster.ca/sedap

Ballantyne, P. (2000). Interpretation of the Arthritis Experience and Decision-making Regarding Treatment: Time, Age and Generational Effects. Ballantyne, Peri. Presentation at the Institute for Human Development, Life Course and Aging, Seminar Series, University of Toronto, April 4.

Berg, B. L. (1995). *Qualitative Research Methods for the Social Sciences*. Needham Heights, MA: Allyn and Bacon.

Chang, R. W., Rellisier, J. M., & Hazen, G. B. (1996). A Cost-effectiveness Analysis of Total Hip Arthroplasty for Osteoarthritis of the Hip. *Journal of the American Medical Association, 275*, 858–865.

Glaser, B. G., & Strauss, A. L. (1967). *The Discovery of Grounded Theory: Strategies for Qualitative Research*. Chicago: Aldine.

Hawker, G. A., Wright, J. G., Coyte, P. C., Williams, J. I., Harvey, B., Glazier, R., Wilkins, A., & Badley, E. M. (2001). Determining the Need for Hip and Knee Arthroplasty – The Role of Clinical Severity and Patients' Preferences. *Medical Care, 39*(3), 1–11.

Hawker, G. A., Wright, J. G., Coyte, P. C., Williams, J. I., Harvey, B., Glazier, R., & Badley, E. M. (2000). Differences Between Men and Women in the Rate of Use of Hip and Knee Arthroplasty. *New England Journal of Medicine, 342*(14), 1016–1022.

Hawker, G. A., Wright, J. G., Coyte, P. C., Paul, J. E., Dittus, R., Croxford, R., Katz, B., Bombardier, C., Heck, D., & Freund, D. (1998). Health Related Quality of Life After Knee Replacement Surgery. Results from the Knee Port Study. *Journal of Bone and Joint Surgery (Amer.) 80-A*(2), 163–173.

Health Canada, Division of Aging and Seniors (2000). Arthritis. Info Sheet for Seniors. Ottawa: Minister of Public Works and Government Services Cat. No. H30-11/8-2E.

Holstein, J. A., & Gubrium, J. F. (1994). Phenomenology, Ethnomethodology and Interpretive Practice. In: N. K. Denzin & Y. S. Lincoln (Eds), *Handbook of Qualitative Research* (pp. 262–272). Thousand Oaks, CA: Sage.

Koskenvuo, M., Daprio, J., Lonnqvist, J., & Sarna, S. (1986). Social Factors and the Gender Difference in Mortality. *Social Science and Medicine, 23*, 605–609.

Liang, M. H., Cullen, K. E., Larson, M. G. et al. (1986). Cost Effectiveness of Total Joint Arthroplasty in Osteoarthritis. *Arthritis and Rheumatology, 29*, 937–943.

Mead, G. H. (1956). *On Social Psychology*. Selected Papers. Chicago. The University of Chicago Press.

Morgan, M. (1980). Marital Status, Health, Illness and Service Use. *Social Science and Medicine, 14*, 633–643.

O'Rourke, N., & Wenaus, C. (1998). Marital Aggrandizement as a Mediator of Burden Among Spouses of Suspected Dementia Patients. *Canadian Journal on Aging, 17*(4), 384–400.

Peyron, J. G., & Altman, R. D. (1992). The epidemiology of osteoarthritis. In: R. W. Moskowitz, D. S. Howell, V. M. Goldberg & H. J. Mankin (Eds), *Osteoarthritis: Diagnosis and Medical/Surgical Management* (2nd ed., pp. 15–37). Philadelphia: WB Saunders.

Qualis Research (1998). The Ethnograph v5.0 for Windows. Distributed by Scolari Sage Publications Software, London, U.K.

Rose, H., & Bruce, E. (1995). Mutual Care but Differential Esteem: Caring Between Older Couples. In: S. Arber & J. Ginn (Eds), *Connecting Gender and Ageing. A Sociological Approach* (pp. 114–128). Buckingham, U.K.: Open University Press.

Strauss, A. L., & Corbin, J. (1994). Grounded Theory Methodology: An Overview. In: N. K. Denzin & Y. S. Lincoln (Eds), *Handbook of Qualitative Research*. Thousand Oaks, CA: Sage.

Susser, M. (1973). *Causal Thinking in the Health Sciences: Concepts and Strategies of Epidemiology*. New York: Oxford.

Trovato, F., & Lauris, G. (1989). Marital Status and Mortality in Canada. *Journal of Marriage and the Family, 51*, 907–922.

Verbrugge, L. M., Lepkowski, J. M., & Imanaka, Y. (1991). Levels of Disability Among U.S. Adults with Arthritis. *Journal of Gerontology, 46*, S71–S83.

Wyke, S., & Ford, G. (1992). Competing Explanations for Associations Between Marital Status and Health. *Social Science and Medicine, 34*(5), 523–532.